Palgrave Historical Studies in the Criminal Corpse and its Afterlife

Series editors
Owen Davies
University of Hertfordshire
Hatfield, UK

Elizabeth T. Hurren
School of Historical Studies
University of Leicester
Leicester, UK

Sarah Tarlow
History and Archaeology
University of Leicester
Leicester, UK

This limited, finite series is based on the substantive outputs from a major, multi-disciplinary research project funded by the Wellcome Trust, investigating the meanings, treatment, and uses of the criminal corpse in Britain. It is a vehicle for methodological and substantive advances in approaches to the wider history of the body. Focussing on the period between the late seventeenth and the mid-nineteenth centuries as a crucial period in the formation and transformation of beliefs about the body, the series explores how the criminal body had a prominent presence in popular culture as well as science, civic life and medico-legal activity. It is historically significant as the site of overlapping and sometimes contradictory understandings between scientific anatomy, criminal justice, popular medicine, and social geography.

More information about this series at
http://www.springer.com/series/14694

Floris Tomasini

Remembering and Disremembering the Dead

Posthumous Punishment, Harm and Redemption over Time

Floris Tomasini
University of Leicester
Leicester, UK

Palgrave Historical Studies in the Criminal Corpse and its Afterlife
ISBN 978-1-137-53827-7   ISBN 978-1-137-53828-4 (eBook)
DOI 10.1057/978-1-137-53828-4

Library of Congress Control Number: 2017938625

© The Editor(s) (if applicable) and The Author(s) 2017. This book is an open access publication
The author(s) has/have asserted their right(s) to be identified as the author(s) of this work in accordance with the Copyright, Designs and Patents Act 1988.
Open Access This book is licensed under the terms of the Creative Commons Attribution 4.0 International License (http://creativecommons.org/licenses/by/4.0/), which permits use, sharing, adaptation, distribution and reproduction in any medium or format, as long as you give appropriate credit to the original author(s) and the source, provide a link to the Creative Commons license and indicate if changes were made.
The images or other third party material in this book are included in the book's Creative Commons license, unless indicated otherwise in a credit line to the material. If material is not included in the book's Creative Commons license and your intended use is not permitted by statutory regulation or exceeds the permitted use, you will need to obtain permission directly from the copyright holder.
The use of general descriptive names, registered names, trademarks, service marks, etc. in this publication does not imply, even in the absence of a specific statement, that such names are exempt from the relevant protective laws and regulations and therefore free for general use.
The publisher, the authors and the editors are safe to assume that the advice and information in this book are believed to be true and accurate at the date of publication. Neither the publisher nor the authors or the editors give a warranty, express or implied, with respect to the material contained herein or for any errors or omissions that may have been made. The publisher remains neutral with regard to jurisdictional claims in published maps and institutional affiliations.

Cover illustration: © Melisa Hasan

Printed on acid-free paper

This Palgrave Pivot imprint is published by Springer Nature
The registered company is Macmillan Publishers Ltd.
The registered company address is: The Campus, 4 Crinan Street, London, N1 9XW, United Kingdom

Preface

In the main, this book is a cross-fertilisation of history and philosophy in the broadest possible sense. Ideas of death, posthumous harm, punishment and redemption germinate in both the conceptual ground of philosophical analysis and the empirical ground of historical case study.

One metaphor for the approach to this book is of germination. At the beginning of this project I had more of a conceptual understanding of death and harm than any sustained appreciation of how that related to any historical contextualisation. However, after spending three years with archaeologists and historians my conceptual training as a philosopher germinated in an appreciation of historical case studies.

Cross-pollination of ideas from history to philosophy chapters and vice versa has led to the germination of an interdisciplinary perspective. That said, each of the chapters also stands alone as either broadly philosophical or historical. This makes the character of the work also appear multidisciplinary in nature.

I cannot make any claims to personally partaking in any serious historical research. I have not visited archives, unearthed any undiscovered and illuminating primary sources. Instead I have stood on the broad shoulders of historians who have inspired me to refashion the lens of philosophical/conceptual inquiry.

Another metaphor that can be used to understand my approach is the fashioning of a varifocal interdisciplinary lens.

So, what is the difference between an ordinary disciplinary lens and a varifocal interdisciplinary lens? A disciplinary lens is pre-ground with a fixed focal length that illuminates a particular kind of academic territory. The problem with importing a disciplinary lens to a different academic territory altogether is that while it illuminates some things in an unexpected way, it will also distort many other details. I began my foray into a historical case study of posthumous harm and redemption in this way.

By trying to understand the conceptual distortions as well as the illuminated focus, I could sense, not re-grind/grind in order to widen the field of view, steadily increasing over-all illumination and the ability to focus near and far. This gave my lens a varifocal quality: a way of looking both out to the far distance of key conceptual distinctions and the near distance, which added sharp empirical focus to a conceptually informed history.

I would like to thank the criminal corpse team at the School of Archaeology and Ancient History, University of Leicester for providing a philosopher with a different way of seeing. My thanks go out to: Prof. Owen Davies, Prof. Pete King, Prof. Elizabeth Hurren, Dr. Emma Battell Lowman, Dr. Rachel Bennett, Dr. Francesca Matteoni, Dr. Shane McCorristone and Dr. Richard Ward. I would also like to acknowledge all the historians and philosophers who have deeply influenced this work and who appear more formally in the references.

A special thank you goes out to philosopher and historian Jonathan Reé, the archaeologist Sarah Tarlow and the poet Sarah Hymas. Jonathan Reé was encouraging and inspiring in equal measure. Sarah Tarlow believed in this kind of project from the start and was instrumental in bringing in a philosopher with interdisciplinary interests. Sarah Hymas patiently read through the draft with an editorial eye.

Last but not least I would like to thank our funders the Wellcome Trust for funding 'Harnessing the power of the criminal corpse' project (Grant No. 095904/Z/11/Z). Without the financial support of a Wellcome Trust Research Fellowship this project is unlikely to have ever materialised.

Leicester, UK   Floris Tomasini

Contents

1 Introduction   1

Part I Conceptual Groundworks

2 What and When Is Death?   7

3 Posthumous Harm, Punishment and Redemption   21

Part II Historical Case Studies

4 Capital Punishment, Posthumous Punishment and Pardon   41

5 Posthumous Harm in the History of Medicine   73

Index   99

CHAPTER 1

Introduction

Abstract The introduction summarises a chapter outline and discusses some of the main concepts used.

Keywords Biological and social death · Posthumous harm
Punishment and redemption

Part I of this book—the conceptual groundworks—is philosophically orientated in character. It consists of two themes: a reflection on what and when is death (Chap. 2), followed by a discussion on the possibility of posthumous harm, punishment and redemption (Chap. 3).

Chapter 2 theorises death as a form of change. Biologically speaking, death is a complex change where it is already present as part of the dying process. Death as a process (dying) can be contrasted with death as a state: dead *or* alive. This work broadens out what we mean by death and its timing. In doing so, it is worth distinguishing social death from biological death.

Social death is understood as a series of narrative changes to the identity of a person that happen as a consequence of real changes to their biology. In absolute terms, social death involves the extinction of someone's biographical narrative. For example, narrative death in an absolute sense is not only no longer being remembered, but also being extinguished from memory and the historical record altogether.

Social death is also intelligible as a processual change, where significant narrative changes configure and refigure personal identity before

© The Author(s) 2017
F. Tomasini, *Remembering and Disremembering the Dead*,
Palgrave Historical Studies in the Criminal Corpse and its Afterlife,
DOI 10.1057/978-1-137-53828-4_1

and after biological death has taken place. As such, social death is not necessarily co-terminus with death as a biological event. For example, the social death of a person may happen as a result of brain injury, so that while an individual may physically survive a major brain injury, they may no longer be the same person. Indeed, certain brain injuries may lead to the 'autobiographical' death of a person, as in the case of those who are left in a permanent vegetative state.

Chapter 3 attempts to understand the harm, punishment and redemption of death. If the ontological and epistemological puzzle of what and when death is becomes the subject of Chap. 2, its normative sense is explored in Chap. 3. When death is has a normative as well as ontological/epistemological dimension. As well as asking the question what and when is death, we can ask the question: what constitutes the harm of death?

Chapter 3 argues that while it is impossible to physically harm or save the dead, it is possible to harm or redeem how we remember them. Looking deeply into the notion of harm, it is possible to distinguish intrinsic from symbolic harm in order to clarify what is meant by this. We can only intrinsically harm human beings that are still living. In this sense it is impossible to harm a corpse. This said, it is possible to symbolically or narratively harm the dead. We can harm:

- the narrative and fidelity of memory;
- the biography of a person that once existed;
- the memory of a person that once existed even though they are still physically alive;
- the memory of a person that no longer exists;
- the symbolic unity of the corpse, whereby dismembering the corpse affects being able to remember the person as they were.

Harming the dead in this way is understood as disremembering the dead, whereas faithfully remembering the dead, as they really were in life, implies redemption. It is this play of words that flags up the significance of the title of this volume. Furthermore, it is possible to conceptually distinguish types of posthumous harm and redemption, which is explored more deeply in Part II.

Part II of this book involves historical case studies—where the conceptual groundworks in Part I cross-pollinate and fertilise in a critical examination of carefully selected case studies where ideas of posthumous punishment, harm and pardoning (redemption) are examined in their historical context.

Posthumous punishment involves retribution, by which the narrative of those that once existed is intentionally and deliberately harmed by institution or state. In its most virulent form, posthumous punishment involves a double form of retributive justice in the eighteenth century: hanging criminals (capital punishment) and dismembering the criminal corpse after hanging either by dissection or gibbetting (posthumous punishment). In its less virulent form, posthumous punishment in the twentieth century involves dishonouring the dead without dismemberment. For example, those deliberately executed by firing squad for a variety of military offences in the First World War were intentionally dishonoured as an example to others.

Chapter 4 opens with an examination of capital punishment through the lens of the British Army's 'shot-at-dawn' policy during the First World War. This leads into a historical discussion of the character of retributive justice and posthumous dishonour of those executed by firing squad, and whether or not posthumously pardoning those shot at dawn is at all appropriate today.

If it is possible to symbolically harm the dead, it is also possible to symbolically redeem their memory. This is intelligible in terms of posthumously pardoning those that were punished and dishonoured.

Some historians argue that posthumous pardoning is either unintelligible and or inappropriate because it is an attempt to re-write history. Such historians have not given enough thought as to what a posthumous pardon is good for. Indeed, it is argued that it is perfectly possible, as well as morally appropriate, to re-evaluate the past in the present for good reason; for example, by rehabilitating the identity of those that have been historically dishonoured in the memory of those still living today.

The chapter ends with a long view of the history of capital punishment, posthumous punishment and redemption, examining how these notions have repeated with a difference over time.

Chapter 5 looks into the idea of posthumous harm in the context of the improper removal and retention of children's organs and tissues at Alder Hey Children's Hospital in the 1990s. Posthumous harm in this historical context is understood as a failure of institutional trust, where moral blindness to inappropriate post-mortem practices thrived in the late twentieth century. While the *effect* may be very similar to posthumous punishment in earlier times, the intention and context are not comparable. The supposed intention behind the inappropriate removal and retention of tissues and organs from dead children at Alder Hey was ostensibly motivated to save lives through medical research, even though in reality the institution colluded in perpetrating harm.

The Alder Hey scandal concerns a failure in a system of trust, where clinicians and their managers were wilfully blind to parental anger and grief brought about by inappropriate removal and retention of their dead children's organs. By outlining two different contexts of understanding Alder Hey and posthumous harm, there is an attempt to provide conceptual clarity as to why posthumous harm mattered at both an institutional level of trust and at a personal level of grief.

To end, there is an attempt at a historical long-view, where 'organ-snatching' at Alder Hey is a practice that has certain similarities with, as well as important differences to, 'body-snatching' in the Georgian period.

Open Access This chapter is licensed under the terms of the Creative Commons Attribution 4.0 International License (http://creativecommons.org/licenses/by/4.0/), which permits use, sharing, adaptation, distribution and reproduction in any medium or format, as long as you give appropriate credit to the original author(s) and the source, provide a link to the Creative Commons license and indicate if changes were made.

The images or other third party material in this chapter are included in the chapter's Creative Commons license, unless indicated otherwise in a credit line to the material. If material is not included in the chapter's Creative Commons license and your intended use is not permitted by statutory regulation or exceeds the permitted use, you will need to obtain permission directly from the copyright holder.

PART I

Conceptual Groundworks

'There are three deaths. The first is when the body ceases to function. The second is when the body is consigned to the grave. The third is that moment, sometime in the future, when your name is spoken for the last time.'
David Eagleman, 2010, p. 23.

CHAPTER 2

What and When Is Death?

Abstract This chapter is one of two conceptual chapters that set up the analytical foundation for the remaining empirical case studies which are mainly historical in character. The first chapter focuses on the question: what is death? The secondary question: when death occurs, depends on what we think death is. This chapter addresses a number of questions: What and when is biological death? Can biological death be understood as an absolute state and/or is it partially present in the process of dying? What is social death? When is social death co-terminus with biological death? When is it not? How can we characterise the meaningful similarities and differences between biological and social death? Why should this matter?

Keywords Biological and social death · Real and symbolic change

BIOLOGICAL DEATH

The commonplace notion of death is to characterise it as an end state: being dead. Nevertheless being dead is not the same as the event of death or the dying process (Scarre 2007, p. 5). Biological death can be understood as:

1. A final event.
2. An absolute state (being dead).

3. Part of the dying process.

Defining Death

The absolute state of being dead is synonymous with the idea of medical death. The definition of being dead, as proposed by the US President's Commission for the Study of Ethical Problems in Medicine and Biomedical and Behavioural Research set up by Ronald Reagan (1981), is when:

> ...an individual who has sustained either (1) irreversible cessation of circulatory and respiratory functions, or (2) irreversible cessation of all functions of the entire brain, including the brain stem, is dead. (Leming and Dickinson 2002, p. 43; Scarre 2007, p. 6)

Death: Absolute State, Final Event and Process

The difficulty with the above definition is capturing the irreversible final moment of death. It is worth critically interrogating both clauses of the above definition.

Clause (1) does not accurately capture the timing of the final biological death event. That is to say, irreversible and irreparable damage to heart and lungs will quickly and inevitably lead to entire brain death, but it is not quite synonymous with that final event. There is a time interval in which the brain is dying because of lack of a supply of oxygen-rich blood to keep it alive, at which point the human brain is dying but not yet dead (Scarre 2007, p. 6).

Clause (2) points to the timing of the final event. The certitude around entire (whole) brain death follows from a clinical assessment of total brain failure. However, the assessment of total brain failure has courted controversy.

The neurologist Alan Shewmon is a leading critic of equating total brain failure with human death. Shewmon identified many cases of patients who were diagnosed with total brain failure that nevertheless ended up surviving. Shewmon collected 175 case reports of patients that had survived against the odds, and whose bodies had stabilised long after the period accounted for by current literature on 'brain death'. The length of patient survival varied from a month to a year and even, in the exceptional Florida Boy Case, 14 years (Rubenstein 2009, pp. 37–38).

In certain cases, therefore, it may be possible to try to artificially sustain a body after so-called total brain failure has been diagnosed. As such, it is possible to distinguish total brain failure from chronic brain death.

Shewmon's arguments have thrown significant doubt over associating death with total brain failure.

This is illustrated in the famous Florida Boy Case. The boy survived for 14 years in an Intensive Care Unit (ICU) after an initial diagnosis of total brain failure. Following his parents' wishes, the boy was artificially ventilated, fed and hydrated in hospital, by which time his body had grown, recovered from wounds and even parts of his brain had become replaced 'by ghost-like tissues' (McMahan 2002, cited in Scarre 2007, p. 7).

The Florida Boy Case has shown that establishing death may be less about precise diagnosis of the brain state and more about understanding the resilience of the human organism as a whole. In other words, the Florida Boy's resilience was tied up with what Shewmon calls the organism's ability to function as an 'emergent property of the whole' (Rubenstein 2009, p. 38). This fits with what Aristotle calls '*entelelchia*', his ancient term for the soul, which has biological connotations with what Joe Sachs has translated as the organism 'being-at-work-to-stay-itself' (cited in Rubenstein 2009, p. 41).

Chronic brain death, where a patient may continue to exist in a permanent vegetative state (PVS), is a notion that only shows up as mattering in the highly advanced technical environment of ICU where specialist clinicians can artificially hold medical death at bay. Arguably then a diagnosis of total brain failure (or indeed chronic brain failure) is:

> ...perfectly correlated with the permanent cessation of functioning of the organism as a whole because the brain is necessary for the functioning of the organism as a whole. It integrates, generates, interrelates, and controls complex bodily activities. A patient on a ventilator with a totally destroyed brain is merely a group of artificially maintained subsystems since the organism as whole has ceased to function. (Bernat, cited in Rubenstein 2009, p. 36)

To conclude, 'life' after extensive brain death is an ambiguous state, one where precise terms are necessary to establish what exactly a human life is constitutive of.

A philosopher that is clear about what *bare* life entails is Leon Kass. He describes life at its most basic as a 'series of preconscious needs.' From Leon Kass's book *The Hungry Soul* (1994):

> What moves an organism to feed is not merely the sensed and registered presence or absence of a certain chemical or edible being in its environment but the *inner needy state* of the organism, for which such an absence is a lack, an absence to be overcome and remedied... The organism would not 'respond' to perceived food 'stimuli' were it not ... 'appetitive' being ... internally ordered toward the necessary activities of self-nourishing. (Leon Kass, cited in Rubenstein 2009, p. 43)

As the Florida Boy Case illustrates, the organism as a whole retains a preconscious and 'inner needy state' for basic appetitive functions. That is, the need for air, hydration and nutrition. This inner state of neediness is met at the threshold of life in ICU, where the organism is not only maintained but even grows, adding to the illusion of recovery.

What and when is death here? It depends on one's perspective of life.

From an understanding of *bare* life, the Florida Boy was a biologically living, growing organism with pre-conscious needs and an inner needy state. From the perspective of a living *person*, the Florida Boy is likely to have died well before his parents projected their hope on to his recovery.

To elucidate further, patients in the UK, who remain comatose and unresponsive and who have made no significant recovery after 12 months from a serious brain injury of this sort, are categorised as being in permanent vegetative state with a statistically improbable chance of recovery (http://patient.info/doctor/vegetative-states).

What is surprising in the Florida Boy Case is how he survived in ICU for 14 years. The ambiguity of his state of existence was probably obscured within the ICU environ. Steps may have been taken to establish how he may have fared without a ventilator, establishing whether or not the boy's brain had the necessary integrative function to sustain autonomous biological life beyond life support. This throws up another distinction: between *bare* life in the technological setting of intensive care and the *bare* life of a deeply brain-damaged individual who may survive for years afterwards with constant care from family and social care professionals.

In the case of bare life the patient can be described as *already* being in a state of 'techno-death', where machines, like ventilators, take over from biological sub-systems that have permanently and irreversibly failed.

Some thinkers regard the neurological standard of whole-brain death to be unnecessarily restrictive (e.g., Green and Wikler 1980). Even if a body could survive technologically unaided, 'neocortical' (or 'higher' brain) death may have occurred anyway, meaning that what remains is a severely mentally and physically disabled individual whose personhood is barely recognisable.

Personhood is characterised by having the mental capacity to be self-aware, communicate with others, and self-create a meaningful life. Once that is gone it is difficult to relate to that human being in the same way. The person who one may once have known has died, presenting the challenge of forming an altogether different relationship with another being. Again the Florida Boy provides an example: while his *autobiographical* life as a *person* was over, destroying who he had been, his *biographical* life was sustained through the narratives of hope his family harboured in his recovery.

Death as Change—A Historical Long-View

The conundrum of understanding biological death is not a new one. It has a long historical root. This is evident in how medical men of the past understood death as both a *state* and a *process*.

Hurren (2013a,b) reminds us of the work of Dr Philips. Dr Philips, in a paper given to the Royal Society in 1834 called *The Nature of Death* describes death in two ways: 'the name of death' where 'sensorial, nervous and muscular systems' were in the process of shutting down. This is roughly equivalent to what we may understand today as a 'living death', inimitable within the process of dying. Philips contrasted this process with a permanent physiological shut down or 'absolute death' (e.g., Philips 1834, cited in Hurren 2013a,b).

Moreover, the idea that dying was sometimes reversible was demonstrated through very early resuscitation techniques. Indeed as early as the 1760s, there were mechanical ways to resuscitate dying persons through artificial respiration in the case of drowning. By 1796 the London Humane Society, for example, claimed to have resuscitated over 2000 people (Hurren 2013a).

Our understanding of the state and process of death has greatly evolved, partly as result of a more sophisticated understanding of brain death in the twentieth century, and partly as result of more advanced resuscitation techniques pioneered by Peter Safar's ABC of cardio-pulmonary resuscitation (CPR), which are now standard practice in emergency medicine (Acierno and Worrell 2007).

Today we have a nuanced understanding of the process of dying, which in its crudest form may be subdivided into roughly six categories:

- Reversible and natural. For example, death may be part of the natural cycle of regenerating the body;
- Irreversible and natural. Death, for example, is part of ageing;
- Reversible and catastrophic. Having a cardiac arrest is reversible, in that the patient can be resuscitated. At this point the patient may be described as clinically but not medically dead;
- Irreversible, catastrophic and unambiguously fatal. Total brain failure that is not redeemable in an ICU environment and is characterised as medical death;
- Irreversible, catastrophic and survivable if technologically aided. Serious brain injury may not necessarily be fatal—persons affected by serious brain injuries survive and sometimes make remarkable recoveries in ICU;
- Irreversible, catastrophic and survivable if technologically unaided. Survivors of major brain injuries that eventually make it out of ICU may be severely mentally and physically disabled requiring life-long support and care. Those who survive the initial crisis and are eventually discharged from ICU and hospital care may have personalities that are barely unrecognisable from before.

A More Conceptual View of Death

On a more conceptual level, death may be theorised in the following ways:

- as a form of change;
- as a particular kind of personal identity.

Death as Change
Geatch (1969) distinguishes between *Oxford* and *Cambridge* changes, characterising Oxford changes as real changes in the intrinsic nature of things and Cambridge changes as relational changes that happen as a consequence of real changes (e.g., Lowe 2002).

Death therefore, takes on a dual aspect: a biological and social aspect. If biological death can be understood as a *real* change in the intrinsic matter of biology, then social death, by contrast, is a relational or narrative change that happens as consequence of real changes in the intrinsic nature of biological materiality.

So, if Maud suffers a brain injury and she is left in a permanent vegetative state, then as a consequence of real changes in the intrinsic property of her brain she will have undergone an irreversible form of biological death. Now a real change in the intrinsic integrity of Maud's brain will result in a relational or narrative change in who we understand Maud to be after her brain injury. Maud might be in a permanent vegetative state (PVS), in which her brain that is responsible for her personality has died before the rest of her body has.

So, implicit in the so-called scientifically neutral language of intrinsic changes in biological properties of her brain, there is also a 'narrative' understanding about who remains. In this way social death is already inextricably linked with biological death.

Social Death

Social death is a relational or narrative change in the meaning of a human life. It involves a change in the narrative identity of persons that either *still* exist or have *once* existed.

Narrative Identity

One way of conceptually fleshing out the difference between social and biological death is to think through two senses of personal identity.

Paul Ricoeur (1992) reminds us that Latin has two meanings for the word identity: identity understood as 'being the same' (*idem*), usually interpreted as the question 'what am I?'; and 'oneself as the self-same' or 'self-constancy' (*ipse*), understood in the question 'who am I?' (e.g., Simms 2003, p. 102).

Now biological death primarily concerns idem identity, where death marks a real change in the intrinsic properties of 'being the same' biologically.

Moreover, the death of 'what I am' (idem) is inextricably linked to being able to self-configure the story of one's life. In short, the physical end of 'what I am' as a living person spells a particular kind of social death: the autobiographical death of one's narrative identity.

Traditional definitions of medical death are unambiguous, describing a final event that leads to the absolute state of being dead—in which case the biological death of a human being (idem identity) is co-terminus with the social death of the person (ipse identity).

The biological death of a person has narrative consequences in how we may configure personal identity. In the most formal terms this involves correct signification. Being dead signifies a corpse, a state of non-being, for which the personal pronoun in the phrase 'I am (a corpse)' is no longer correct. A corpse refers to a husk, and a husk is no longer a person that actively possesses a body. Furthermore, physical death has relational consequences for others. My death, for example, would mean that my wife would undergo a relational/narrative change: that is, my wife would become a widow.

Social death concerns our ipse identity—the narrative identity of who we are. While social death is dependent on having existed, it is not necessarily co-terminus with existing as a biological entity.

Real changes in our biology certainly prompt relational changes in how we may configure the narratives of our lives. After a heart attack, for example, there may be a subtle shift in who we understand ourselves to be through what we believe we are realistically capable of doing. This may signal a subtle shift in our personality. Less subtly, brain injury, as argued earlier, can lead to narrative inversions in our clinical status: from a living person that is self-conscious and aware of others, to a living human organism which in PVS is not conscious in this way.

A relational change in the meaning of who a person is has both an existential as well as a biological dimension. The narrative of who we are is *existentially anticipated* in the face of our physical mortality. We existentially configure the meaning of our lives in anticipation of our physical death. This has a secular and religious dimension.

In secular terms the meaning of our life matters beyond its physical annihilation. For example: we might suffer a serious brain injury that marks our autobiographical death destroying our dignity; we resonably anticipate being respectfully treated and honestly remembered after our physical life has ended.

In spiritual/religious terms we may anticipate who we are and how we might continue in a life hereafter. If one believes that how we treat mortal remains matters for a disembodied life hereafter, then the burial rituals associated with keeping the corpse intact take on a special 'narrative' significance.

The narrative identity of the dead is also *refigured* by others who survive them. The narrative identity of the dead is refigured for good and ill through memory and biography by those who survive the deceased. Informally, we are remembered by family and friends, while formally we may be remembered through the social impact of our past actions. In short, who we are survives our physical death in narratives of remembrance.

Similarity and Difference: Biological Versus Social Death

There are both fundamental similarities as well as underlying differences when comparing social with biological death.

- Biological and social death both involve change.

However, how we understand 'change' differs in each case. Biological death involves understanding *intrinsic* or *real* changes to the material what-ness of existence, whereas social death involves an understanding of *narrative* change of who we are.

- Biological and social death depend on an existence condition. Death is unintelligible without ever having existed, or never having changed.

How we interpret the existence condition is different depending on context; that is, it has different meanings in the context of biological and social death. Biological existence primarily concerns changes in the material what-ness of life. In the case of social death however, existence primarily concerns a narrative self-constancy of who-ness over time.

Narrative identity is *not* necessarily co-terminus with physical existence; that is, a human being, either existentially anticipates and configures the meaning of who they are beyond their physical annihilation, or their narrative identity changes through its refiguration by others who survive them.

- Biological and social death can be understood as a state or a process.

Being dead is a physical state to be contrasted with dying, which is a biological process. In social terms, the absolute state of being dead is the equivalent to being completely forgotten and expunged from the historical record altogether. Moreover, the social equivalent of physically dying is being slowly forgotten, misremembered and, most damagingly of all, being harmfully remembered (disremembered). In this volume the focus is on the process of being disremembered, which is a form of social death.

- Understanding when biological death is socially contested.

Biological death is simply a value neutral intrinsic change in a person's biology. How one interprets this, however, is socially contested: what biologically dies is inextricably linked to how we interpret and value the existence of who it is that dies.

In an effort to establish formal legal clarity, standard definitions attempt to make biological death as straightforward as possible, by making it an end to an individual's life and narrative. For example, the standard definition of a person's death is of 'an individual who has sustained either... irreversible cessation of circulatory and respiratory function, or irreversible cessation of all functions of the entire brain...' (e.g., see the first page of this chapter). This is incorporated into current US law, under the Uniform Determination of Death Act (UDDA 1980) and signifies being unequivocally dead.

However, in the UK legal precedents pre-empt the social death of a human being before physical biological life has actually ended. Take the example of Tony Bland.

Anthony Bland was allowed by the high court to be medically euthanised, setting a precedent for passive non-voluntary euthanasia in UK hospitals. Bland was fatally injured at the Hillsborough football stadium in 1989, developing serious injuries in the crush at the Leppings Lane terrace that caused an interruption to the supply of oxygen to his brain. As a result he was left with irreversible brain damage to the higher centres of the brain that support personhood, but the brain stem function was left intact. In short, whilst Anthony Bland wasn't strictly biologically/medically dead, the person who was Anthony Bland was dead. This led his family to petition the high court to carry out a form of passive non-voluntary euthanasia; that is, euthanising Tony Bland by withdrawing artificial nutrition and hydration. The high court ruled in favour of the family, acknowledging that it was highly improbable that Bland would ever emerge from his persistent vegetative state (known today as 'permanent vegetative state'). The court judged that passive euthanasia was morally and legally permissible. This, the court judged was because 'to his parents Tony Bland was dead,' and in PVS, his 'life was of no benefit to him as a person' (cited in Robertson 1996, p. 723–746). In other words, the social death of Anthony Bland became the deciding factor in how to respond to the bare biological life that remained. So, in order to *redeem* the memory and dignity of who he was to his parents, it was necessary to prematurely end it.

- Only social death has a normative valence.

Biological death is an event, a fact. It carries no moral valence in and of itself. It is either a final event or it is a significant change that is already happening as part of a process of dying. How we socially interpret biological death as an event, however, is far from neutral. Biological death is narratively interpreted as 'good' or 'bad'. This is because the meaning of social death has a moral valence that we project onto the physical event by way of the expectations we weave into how biological life has existential meaning for us. We might describe a 'good' death as one by natural causes, whereby a person dies peaceably in the fullness of time. This can be contrasted to the typical narrative of a 'bad' death, where a person might 'suffer' because they have a disease and choose to 'fight' for life regardless and 'rage, rage against the dying of the light' (Thomas 1971).

The normative valence of death as an existential challenge is not only narratively configured *before* the biological event actually takes place, it is also narratively refigured in how we are remembered by others. We hope to be well remembered or at least honestly remembered. This refiguration of our narrative existence is considered good and just. We hope not to be misremembered and certainly don't want to be disremembered. This refiguration of who we were is considered bad and unjust.

- Social death is subject to redemption.

Death is a biological fact of the necessary impermanence of human life. In the neutral scientific language of biological changes, death may or may not be temporarily reversed, but cannot be avoided. Yet the language of medicine is laced with heroic and redemptive metaphors: doctors 'cure' and 'save lives' and are involved in medical 'breakthroughs' against 'killer' diseases.

The fight to 'save life' is a redemptive narrative that we project onto what is simply a value neutral biological event, where dying, biologically speaking, is either, catastrophic and irreversible, or not. In other words, there is little sense in talking about redemption in terms of death as a biological event that we will all succumb to, unless we have already added value to what it means for us to live rather than die.

The redemptive narrative of who a person was and what that means continues on after a person has ceased to physically exist. This narrative of remembering is rarely straightforward, because it involves a reconstruction of a person's posthumous identity, which again adds an interpretive meaning about the value of a life lived. This may or may not be broadly resonant with how that person tried to live their life. If that refigurative narrative is true to the facts and spirit of a life lived, then there is a tendency to think that the deceased is well-remembered. If that posthumous narrative is deliberately harmed after they have died we tend to think of the deceased as having been disremembered.

Being historically disremembered, by being perceived as a notorious historical figure for example, may be time limited. In other words, it is possible to change our minds about whether historical figures deserve the self-same notoriety today. If this is the case then posthumous pardoning may be necessary. This, it is argued, is not so much about rewriting the past, but about re-evaluating the past in the present, where a 'new' narrative of forgiveness may, for moral reasons, legitimately exist alongside the notoriety of historic ones.

The Harm and Redemption of Death

The author's main interest in the remainder of the short volume is in social rather than biological death. What follows is an examination of how narrative identity is subject to posthumous harm, punishment and redemption. In carrying this out, there is further conceptual examination of the possibility of posthumous harm and redemption.

Summary

This second chapter has discussed what and when death is, conceptualising biological and social death both as a *state* and a *process*. The understanding of biological death as a process complicates what and when we understand death to be.

Death has been theorised two ways, as two forms change and personal identity. While the two kinds of death are certainly related, social death is not necessarily co-terminus with biological death. Narrative identity can both be existentially configured in advance of the physical event and/or narratively refigured by others who survive the deceased. Most importantly, social death has a normative valence that physical death as an intrinsic biological event does not.

References

Acierno, L. J., & Worrell, L. T. (2007). Peter Safar: Father of modern cardiopulmonary resuscitation. *Clinical Cardiology, 30*(1), 52–54.

Eagleman, D. (2010). *Sum: forty tales from the afterlives*. USA: Vintage Books.

Geatch, P.T. (1969). *God and the Soul*. London: Routledge.

Green, P. T. & Wikler, D. (1980). Brain death and personal identity. *Philopshy and Public Affairs, 9*(2), 105–133.

Hurren, E. (2013a). The dangerous dead: A murder act, medical death & the criminal corpse in late Georgian England. Unpublished.

Hurren, E. (2013b). The dangerous dead: Dissecting the criminal corpse. *The Lancet, 382*(9889), 302–303.

Kass, L. (1994). *The hungry soul: Eating and the perfecting of our nature*. Chicago: Chicago University Press.

Leming, M. R., & Dickinson, G. E. (2002). *Understanding dying, death and bereavement* (5th ed.). Fort Worth, TX: Harcourt.

Lowe, E. J. (2002). *A survey of metaphysics*. Oxford: Oxford University Press.

McMahan, J. (2002). *The ethics of killing: Killing at the margins of life*. Oxford: Oxford University Press.

Patient. Info. Retrieved August 26, 2016, from http://patient.info/doctor/vegetative-states.

Philips, A. P. W. (1834). On the nature of death. *Philosophical Transactions of the Royal Society of London. 124*, 167–198.

Ricoeur, P. (1992). *Oneself as Another* trans. Blamey, K. Chicago and London: University of Chicago Press.

Robertson, D. (1996). The withdrawal of medical treatment from patients: Fundamental legal issues. *Australian Law Review, 70*(9), 723–746.

Rubenstein, A. (2009). What and when is death. *The New Atlantis, 24,* 29–45.
Scarre, G. (2007). *Death.* Stocksfield: Acumen.
Simms, K. (2003). *Paul Ricoeur.* London: Routledge.
Thomas, D. (1971). *The poems of Dylan Thomas.* Jones, D. (ed.). UK: New Directions.
Uniform Determination of Death Act or UDDA. (1980). American Medical Association and American Bar Association approved.

Open Access This chapter is licensed under the terms of the Creative Commons Attribution 4.0 International License (http://creativecommons.org/licenses/by/4.0/), which permits use, sharing, adaptation, distribution and reproduction in any medium or format, as long as you give appropriate credit to the original author(s) and the source, provide a link to the Creative Commons license and indicate if changes were made.

The images or other third party material in this chapter are included in the chapter's Creative Commons license, unless indicated otherwise in a credit line to the material. If material is not included in the chapter's Creative Commons license and your intended use is not permitted by statutory regulation or exceeds the permitted use, you will need to obtain permission directly from the copyright holder.

CHAPTER 3

Posthumous Harm, Punishment and Redemption

Abstract The idea that we can harm the living is uncontroversial. By comparison, the idea that we can harm the dead is highly controversial. Chapter 3 explores the intelligibility and plausibility of ante-mortem harm, posthumous harm and redemption. This chapter addresses the following questions: Is it possible or impossible to harm the dead? What is ante-mortem harm? How is ante-mortem harm intelligible and how is it not? How might we reframe the harm of death? How might we characterise harms to the dying? What is posthumous harm and how is it intelligible? What is the relationship between dismembering the corpse and disremembering persons? How can we distinguish different kinds of posthumous harm (posthumous harm from posthumous punishment for example)? How might we understand posthumous redemption as a counterpoint to harm?

Keywords Intrinsic · Posthumous harm and punishment · Symbolic and narrative harm

The Impossibility of Posthumous Harm

It is often taken for granted that we can harm the living, but not the dead.

A classical view on the impossibility of posthumous harm comes out of the empirical philosophy of Epicurus.

Epicurus offers us a consolation of philosophy, one that attempts to relieve us of the fear of death. Epicurus's therapeutic philosophy corrects what he sees as philosophical errors that generate an anxiety about an afterlife, or a fear of the event of death itself.

Epicurus was a materialist, subscribing to the atomic theory of the pre-socratic philosopher, Democritus. According to Epicurus 'the human soul is composed of eternal particles which experience sensation only while united with the body' (http://newepicurean.com, sect. 6). He argued that everything, in an infinite amount of combinations, was composed of atoms, including the soul, which he believed to be material. Epicurus relied on empirical evidence of his senses to make judgements and evaluations about the world around him and, through this, reasoned that the soul was the animating materiality that gave locomotion to the body and expression to the features. He believed there was little sense in claiming that the soul was immaterial or incorporeal because 'something that was void could neither act or be acted upon and the soul acts and is acted upon' (http://newepicurean.com, sect. 6). He therefore rejected the view that an immaterial or incorporeal soul existed. This being the case, any anxiety about an afterlife, spurred on by theistic faith, was unnecessary.

Having rejected the immortality thesis (the afterlife of the soul), Epicurus insists that death itself is not something to fear, but an event that we should be indifferent to. In his words:

> Become accustomed to the belief that death is nothing to us. For all good and evil consist in sensation, but death is the deprivation of sensation... So death, the most terrifying of ills, is nothing to us, so long as we exist, death is not with us; but when death comes, then we do not exist. It does not then concern either the living or the dead, since the former it is not, and the latter are no more. (Epicurus 1940, p. 31)

As many writers have noted there are two distinct strands of argument entangled within this succinct position. The first strand relies on what may be called the experience requirement: that is, 'good and evil consists in sensation'. Whereas the second strand is founded on what philosophers have called the existence condition: that is, 'death is nothing to us when we do not exist'.

While it is possible to critically separate the two strands of argument, it is more difficult to reject the argument as a whole. This is because the weaker first strand is inextricably woven and linked with Epicurus' stronger second. So while it is worth picking apart some of the weaker claims Epicurus makes, the Epicurean argument taken as a whole is stubbornly resilient.

The first strand of argument draws on the presupposition that all that matters about death is experience. This is encapsulated in what may be called the experience requirement that is expressed in the second sentence.

The idea that 'all good and evil consists in experienced sensation' is vulnerable to criticism. Nagel, for example, rejects this:

> Most good and ill fortune has as its subject a person identified by his history and his possibilities, rather than merely by his categorical state of the moment – and while the subject can be exactly located in sequence of places and times, the same is not necessarily true of the goods and ills that befall him. (Nagel 1970, p .77)

To illustrate the above, Nagel imagines an intelligent adult who suffers a brain injury. The person who has suffered the brain injury is well cared for and now has the mental condition of a contented infant. Nagel argues that this person has undergone a grave misfortune; not because the person is now on the level of a contented infant, but because the intelligent adult has lost their intelligence. In other words, there are goods or interests that may lie beyond awareness and the immediate experience of being harmed (Nagel, cited in Belliotti 2013, pp. 25–26).

Epicurus also claims that 'death as the deprivation of sensation is neither good nor bad for us' but something that we should treat indifferently.

This prompts an obvious objection: 'once we are dead, while we can no longer suffer *painful* experiences, we can no longer experience *pleasurable* ones either' (Scarre 2007, p. 87). For this reason alone, it is very difficult to treat death indifferently. For those of us who find a positive value in being alive 'any reason to (want to) live is an excellent reason to want not to die...' (Luper-Foy 2002, cited in Scarre 2007, p. 103). Moreover, the converse is also true: there might be good reasons to be dead if life turns out badly. Either way, from the perspective of being thrown into a life that has to be lived in one way or other, it is difficult for death to be nothing to us.

Epicurus goes on to claim '...death, the most terrifying of ills, is nothing to us, so long as we exist death is not with us...'. Epicurus here is presumably running with a binary notion of death: death as an absolute state versus being uncomplicatedly alive. While it is true to say we are not harmed by the *absolute* state of being dead, it is not true to say that death is not with the living when they are in fact dying. To unpack this further: while it may be logical to be indifferent to the state of being dead when there is no subject to experience anything at all, it is equally logical to fear death as a process; to grieve partial loss of our faculties and express its often painful passage.

In sum, Epicurus does not consider the death in dying which can be painful and distressing since it is already busy taking away that which we value most about living. However, while Epicurus either remains silent or unaware of this complication, it is a digression from the spirit of what Epicurus is fundamentally asking us to imagine.

To rephrase from the perspective of being uncomplicatedly alive, perhaps wholly healthy, we can afford to be indifferent about death because we are securely outside of its grip on us. More persuasively still, Epicurus is inviting us to embrace the possibility of the total annihilation of being dead; that is, 'death is nothing to us,' precisely because there is *no subject* that experiences good or bad.

The annihilation of the subject by biological death seems Epicurus's most unassailable strand of argument. It rests on what Fred Feldman (1991) first called the 'Existence Condition' and the proposition that 'nothing either good nor bad can happen to a subject s at time t unless s exists at t' (cited in Scarre 2007, p. 89). To be harmed a person has to first *be*. So, if there is no subject to suffer the harm of death, there can be no harm.

Looking at the problem in the round, there are three major strategies in dealing with the existence condition where absolute death cannot harm us:

- Agree with Epicurus, by following his assumption that the state of being dead can mean nothing to us because we cease to exist as subjects that can be harmed (Partridge 1981);
- Circumvent the existence condition by posing an alternative: *ante-mortem harm*. This allows us to contemplate 'the misfortunes of the dead' (Pitcher 1984) by seeing how posthumous events throw a light on blocked interests of persons that once existed (Feinberg 1984). The Feinberg–Pitcher position has inspired a wealth of conceptual criticisms and some interesting variations on the ante-mortem harm thesis from a wide variety of

contemporary philosophers (Belliotti 2013). This is beyond the scope of this chapter;
- Meet the existence condition head on by reframing posthumous harm and the annihilation thesis. This involves understanding how subjectivity might survive death, and how it might be employed to make posthumous harm an intelligible concept (Scarre 2007; Sperling 2008; Belliotti 2013; Tomasini 2009).

DEATH AND ANTE-MORTEM HARM

The idea that we cannot harm the dead has a long historical root in classical philosophy. Yet it is an idea that has fascinated and vexed contemporary philosophers.

Some philosophers like Ernest Partridge agree with Epicurus, elaborating why the dead are beyond harm:

> Nothing happens to the dead... Accordingly, after death, with the removal of a subject of harms and bearer of interests, it would seem that there can be neither 'harm to' nor 'interests of' the descendent. (Partridge 1981, p. 253)

Pitcher was one of the first philosophers to realise that while Epicurus's argument might be difficult to wholly dismiss, it can be successfully circumvented by posing a different kind of subject to which harms might accrue.

Pitcher accepts the annihilation thesis and asks: 'the dead, if they exist at all, are so much dust. How is it possible for so much dust to be wronged?'(Pitcher 1984, p. 183). In answering his own question he distinguishes between two kinds of subject. The post-mortem subject, call him Harry, perhaps mouldering away in some grave, and the ante-mortem Harry, as he was at some stage of his life before he died. According to Pitcher, while death cannot be bad for Harry's corpse it can be bad for the ante-mortem Harry.

Feinberg agrees with the Pitcher thesis and elaborates on the idea of ante-mortem harm. What follows is philosophical convergence, which brings together Feinberg and Pitcher, and that might for the sake of convenience be referred to as the Feinberg–Pitcher account.

According to Feinberg–Pitcher it is not intelligible to talk of a subject's survival after physical death, nor does it make sense to talk of harming posthumous interests without a surviving interest-bearer. As such they rule out the possibility of posthumous harm.

So how are ante-mortem persons harmed by death? The answer lies in the fact that ante-mortem persons have transcendent interests; interests that can only be fulfilled or thwarted after they are dead. The Feinberg–Pitcher account argues that transcendent interests may be frustrated by death. So: 'the subject of harm in death is the living person ante-mortem, whose interests are squelched' (Feinberg 1984, p. 93).

This conclusion about who is harmed has important consequences for the timing and causation of harm. In Feinberg's own words:

> The ante-mortem person was harmed in being the subject of interests that were going to be defeated whether he knew it or not… It does not become retroactively true that as a subject of doomed interests he is in a harmed state: rather it was true all along… Exactly when did the harmed state of the ante-mortem person, for which the posthumous event is 'responsible', begin…' at the point, well before his death, 'when the person had invested so much in some post-dated outcome that it became one of his interests. (Feinberg 1984, pp. 89–90)

Ante-mortem harm purposely avoids the problem of the objectionable backwards causation (what happens in the future can causally affect what happened in the past). Instead Feinberg is proposing that a person who is harmed ante-mortem was going to be harmed 'all along' by an awaiting event. In a reiteration of Feinberg's words, 'the ante-mortem person was harmed in being the subject of interests that were going to be defeated whether he knew it or not'.

Let us look at a brief biography of Private James Highgate (Watson 2014) who was shot at dawn during the First World War and posthumously dishonoured, to illustrate the point. What follows is a brief timeline of significant events:

- 13 May 1895: Thomas James Highgate is born
- 4 Feb. 1913: Highgate enlists at the age of 17 (and officially commits to a youthful interest in living a life as an honourable soldier)
- 28 July 1914: World War One begins
- 8 Sept. 1914: Highgate is shot at dawn for desertion
- 1920: Private Highgate's name was left off his local war memorial in the village of Shoreham, Kent. Unlike his comrades, his name was omitted.

- 2006: Private Highgate is posthumously pardoned by Labour Government along with 305 other soldiers dishonoured by being 'shot at dawn' for various offences un-becoming of a fighting soldier
- 2014: Private Highgate's name is still missing from the Shoreham war memorial, despite the blanket pardon in 2006.

According to the Feinberg–Pitcher account, when the erection of a war memorial occurred in the village of Shoreham in 1920, the deliberate omission of private Highgate's name could not harm him, as harm is something that only accrues to living persons.

The omission of Thomas Highgate's name from his local war memorial does harm the ante-mortem Highgate's interest in having an unambiguous career as an honourable soldier. This posthumous event (the omission) throws into stark relief that ante-mortem Highgate had been 'playing a losing game' (Feinberg 1984, pp. 91–92) from the moment he developed the interest of being an honourable soldier to the time his name is still missing from the war memorial in 2014.

The Feinberg–Pitcher account is seductive and clever. It skilfully avoids metaphysical pitfalls such as backwards causation, whilst also helping to explain our moral intuitions about how Highgate may have been wronged.

In doing so, the Feinberg–Pitcher thesis might explain how death itself harms ante-mortem interests, without begging the question, how can harm affect non-existent persons? In sum, the Feinberg–Pitcher ante-mortem harm thesis *seems* to solve the tricky existence condition.

However, some telling criticisms remain.

The first criticism stems from the intuition that a significant harm seems to flow from a post-mortem event of his name being omitted from a war memorial in Shoreham village. However, the Feinberg–Pitcher thesis circumvents this intuition because they are still running with the Epicurean annihilation thesis: it is only possible to harm living subjects, not the deceased. Instead of concluding, like an Epicurean might, with a straightforward 'no harm' conclusion, faithful followers of the Feinberg–Pitcher thesis would be forced to argue that ante-mortem Highgate interests were harmed.

By denying the problem of backwards causation, the Feinberg–Pitcher account is vulnerable to yet another causation problem: *phantom causation*. In the historical example, the fact that Highgate's name was left off the Shoreham war memorial is not a cause of harm to Highgate, according to the Feinberg-Pitcher thesis. Feinberg-Pitcher do not take into consideration posthumous harm to someone's reputation, they only consider ante-mortem harm. This is a definite weakness in their account, since it is counter-intuitive not to recognise that there was harm done to Highgate's memory and reputation by ommitting his name from the local war memorial.

The Feinberg-Pitcher thesis has no causal force when it comes to explaining posthumous harms to Highgate. To follow such an ante-mortem thesis requires an appeal to obfuscating language that problematises cause. In other words, the author would need to borrow Feinberg's puzzling language; that is, it is clear that Highgate's interests were harmed 'all along' and that by being left off the war memorial Highgate was 'playing a losing game' in his transcendent interest for honour.

Setting aside the obvious importance of the mortem harms—his trial and execution—it is deeply counter-intuitive to think that leaving Highgate's name off the official war memorial in 1920 was not further harming Highgate in some way.

According to Feinberg–Pitcher we can only squelch transcendent interests of ante-mortem persons. This is unnecessarily restrictive, as it is possible to think of harms in relation to social death as well harms in relation to biological death. In the case of the latter, narrative or ipse identity survives in the form of memory and biography that is refigured by others who survive the deceased. So leaving Highgate's name off the Shoreham war memorial was a symbolic harm to Highgate's memory.

In the author's view it is possible to avoid the problem of phantom causation by broadening out what can be harmed when.

In sum we need to distinguish between the configuration of ante-mortem interests that can be thwarted (Feinberg and Pitcher) and refiguration of a deceased person's memory by others, which, in turn, is also subject to a harm experience by those that survive the dead.

There are two responses to the Feinberg–Pitcher account.

The first is to view it as wrong because it fails to satisfactorily deal with some important timing and causation problems. In my opinion this is uncharitable and mistaken. The Feinberg–Pitcher ante-mortem thesis makes huge strides into conceptualising the harm of death.

The second is to view the argument as an unnecessarily restrictive account of harm, one that does not consider other possibilities. The serious criticism, in the author's view, is that the Feinberg–Pitcher thesis is running with an overly narrow account of annihilation, one that fails to fully consider how personal identity may survive death. This leads to an over-simplistic assignation of categories, dichotomising harm between the possibility of ante-mortem harm on the one hand, and impossibility of posthumous harm on the other.

THE HARM OF DEATH REFRAMED

There seems to be a narrow choice on offer: either agree with Epicurus et al. on the annihilation thesis and the impossibility of the posthumous harm, or circumvent the problem by posing ante-mortem harm (Feinberg–Pitcher). To move beyond this dichotomy we need to think about the problem more broadly and reframe the terms upon which some assumptions about harm and death rest.

The Meaningfulness of Life Beyond Death

The notion that death should be 'nothing to us', as Epicurus puts it, is difficult to reconcile with desires that give meaning to our lives beyond our sense of self-satisfaction. That the meaningfulness of life extends beyond death is testified by our transcendent interests or desires. Or, as Belliotti puts it: 'my biographical life transcends my biological life' (Belliotti 2013, p. 102).

Transcendent desires surpass the timing of our own death and, in a search for meaningfulness outside ourselves, we go beyond the self-satisfaction of meeting our needs and personal preferences in the present moment. Transcendent interests or desires include: the honouring of death bed promises; the disposal of our property and belongings after death; the integrity of our posthumous reputation; the respectful disposal of our corpse according to religious belief and custom; the flourishing of children, relatives and friends; and the successful completion after death of creative endeavours or projects begun in life.

These desires resonate with the human aspiration 'to mould objectively worthy lives' (Belliotti 2013, p. 103). As such they do not flow from internal satisfaction alone. The hedonistic thesis would have it that life is the sum of all worthwhile *individual* preferences satisfied. However we also need our life to mean something beyond itself. Indeed, transcendent desires flow from a sense that a meaningful life has an 'external vantage point' (Belliotti 2013, p. 103), one that contributes to a wider and deeper personal, human and cosmic perspective.

From a personal point of view we care about those we leave behind when we die—children, family and friends. We want to be remembered faithfully, for what our life stood for. Moreover, it is natural that we might bequeath finances, property and belongings to ensure the welfare of significant others. Less tangibly, we believe that our search for meaning is often carried over by our children and their children. In terms of our death, we care that our corpse is respectfully treated—remembered both literally and figuratively after life.

From a wider connection to humanity our life might mean something in respect to creative works we leave behind: uncompleted projects for the wider good of society, fictional, non-fictional works and/or practical projects like scientific endeavours that promote a sense of progress.

Finally, from a cosmic point of view, we may believe that our life means something beyond our material embodied existence. This has ramifications for how we treat the human body when dead. Many people throughout history have been socially conditioned in complex ways by religious and folkloric fears born out of a belief not to 'harm' the dead, lest it hinders safe-passage of the soul in the afterlife.

For those of us who have forgone religious superstition about a disembodied afterlife, it might be important for our lives to have counted for something greater than the sum of its remembered parts. For example: a view that a life lived expresses a solidarity with others, or unity with a natural ecological order of things, all of which might be sensitively reflected in a burial ritual.

Reconsidering the Annihilation Thesis and Existence Condition

One way of assessing the possibility of posthumous harm is to reframe what it is meant by annihilation and existence.

The original formulation of the annihilation thesis is Epicurean: beings cannot be harmed posthumously because at death subjects are annihilated and are thus beyond all experience.

For the sake of this particular argument let us assume, alongside Epicurus, that a life after death is false (although it would be wiser to be agnostic about it). Then Epicurus's assertion that death puts us beyond all experience is likely to be true, as it is impossible to imagine experience without existing in some sense or other.

Epicurus also asserts that all good and evil consist of experience. This assertion, by itself, is difficult to support and much easier to challenge. Epicurus here is running with a highly questionable hedonistic assumption: pleasure and pain define what is good or bad for us. This is true if we understand harm in a very narrow sense of pain or hurt. However, harm may be distinguished from hurt (Belliotti 2013, p. 11). While hurt involves suffering to which we accrue a negative valence through the commonplace expression 'it hurts', harm is something that can happen without necessarily being consciously aware that it is happening.

The classic example that may be used to illustrate this is the case of a peeping Tom spying on our private life. While the peeping Tom may, undiscovered, never hurt us, his actions nevertheless harm a desire for privacy (Belliotti 2013, p. 11). Likewise, while harming a dead body cannot physically hurt the deceased, it can harm the ante-mortem person who holds certain beliefs about how their corpse ought to be treated post-mortem. Whether these beliefs are credible or not is much less important than the fact that they can be harmed by not being respected.

One of the reasons why harms are so much more far reaching than hurts is that they are not constrained by the immediacy of subjective experience and preference satisfaction. As already argued, for a life to be meaningful it often needs to be considered to have a meaning from an external vantage point beyond itself. Harms to the dead outstrip the experiential constraint of hurt for this reason. While it might be absurd to think that we might hurt the dead, it is possible to harm the interests of persons that once existed. It is this posthumous harm, involved in intentionally disremembering someone after they have died, that the Feinberg-Pitcher thesis cannot easily account for.

The Feinberg–Pitcher thesis is right to acknowledge the Epicurean limitation to the annihilation thesis by arguing that the end of the *experiencing* subject does not spell the end of discussion about the misfortunes of the dead. This is certainly an advance on Epicurus and his modern exponents, like Partridge, who do not seem to puzzle through the possibility of harming the interests held by persons that once existed.

Equally the Feinberg–Pitcher thesis is wrong to draw the distinction too tightly between the possibility of ante-mortem harm and impossibility of posthumous harm.

Towards a Typology of Harms

For reasons of analytical clarity it is worth distinguishing between different orders of harm.

- *First order intrinsic harms*, which causally affect living persons and not dead bodies. The dead body, by virtue of being a corpse is beyond experiencing intrinsic harm of any order.
- *First and Second order symbolic harms* are either harms to the interests of those who once existed (first order symbolic harms to ante-mortem persons), or they are biographical harms that occur after death and to a posthumous reputation (second order symbolic harms). The Feinberg–Pitcher thesis considers the first, but not the second. First order symbolic harms accrue to the transcendent interests of ante-mortem persons.

  Second order symbolic harms are harrmful post-mortem re-figurations of what once living persons interests were thought to be. They can be thought of as harms to memory and biography. Posthumous harms, moreover, are not free-floating, they accrue to a living subject bearer, normally in the form of a relative or friend who is concerned that the memory of their dead is not misrepresented. The living are witness to the fidelity of memory and its rightful conservation. They are surrogate interest bearers for the transcendant interests of those now dead.
- *Second order intrinsic harms* are intrinsic harms to significant others who are related to or care about the deceased and experience the hurt of how they are disremembered. They tend to be psychological harms and rely in a narrative investment about how someone ought to be remembered. Being disremembered causes psychological distress in those that invest truth in a certain memorial narrative. Most commonly, it may be experienced by survivors who hear others 'speaking ill' of their dead relatives or friends. More dramatically it may be experienced by a relative as a deliberate and intentional attempt to dishonour the dead and even to refigure their memory as particularly notorious. Usually disremembering the dead is expressed through the spoken or written word when this is not what they deserved.

  Second order intrinsic harms can be experienced by others as harm to the corpse itself. While it is not possible to harm a mouldering corpse *in itself*, the corpse has a symbolic unity for the living that needs to be respected, in particular by those who grieve 'their' dead and experience inappropriate dismemberment as harm to memory and religious belief.

Reviewing and Previewing Harm and Redemption of Dying and Being Dead

What follows is a brief review and preview of some of the responses to some of the standard conceptual assumptions around death and dying, whereby posthumous harm is thought to be unintelligible and nonsensical.

First Assumption: We Are Either Dead or Alive

This view does not take into account death that occurs in the process of dying.

Both the Florida Boy and Tony Bland case in Chap. 1 showed that it is possible to remain technically alive (in a biological sense), but still have undergone a significant social, autobiographical death. Symbolic or narrative death therefore, can occur before biological death has unambiguously ended.

Second Assumption: Ante-Mortem Harm Is Possible, Posthumous Harm Is Not

There is an assumption that in order to circumvent the existence condition, it is possible to talk about ante-mortem harm but not possible to talk about posthumous harm because at this point the subject bearer of harm no longer exists. Again, this is true, but nevertheless flows from an overly narrow view of what it means to exist. For example, it is perfectly intelligible to talk of a person's narrative still existing long after they are dead through memory and biography. Harms to the memory and biography of the dead are second order symbolic harms.

Third Assumption: It Is Possible to Harm a Living Person but Not Their Corpse

This view maybe attributed to many philosophers post-Epicurus who run with the idea that harm necessarily invokes the idea of *intrinsic* harm.

It is by invoking a notion of biological death that we can understand intrinsic harm; only living persons undergo *real* changes that are subject to intrinsic harms. From this perspective, it is nonsensical to talk about harming an ex-person's ashes or their mouldering corpse.

While this is all true, it does not take into account the idea of social death—where relational changes happen as a consequence of real or intrinsic changes involved in biological death. In other words, while it may not be possible to intrinsically harm the dead body, it is possible to symbolically harm the memory and biography of the dead persons that once existed through inappropriate dismemberment. In Chap. 5 the symbolic harm to memory is explored through complicated grief. Relatives who had been subject to the shock of hearing that their dead children had been dismembered through inappropriate post-mortem procedures at Alder Hey in the 1990's, had difficulty remembering their children as they were in life. This involves both second order symbolic and intrinsic harm. Symbolic because it involves descration of the corpse and what this represents and intrinsic because it is psychologically distressing to experience this.

On another level, dismembering the dead is an assault on cultural and spiritual beliefs about how the dead should be treated after life. For example, in the eighteenth century, criminals who had been sentenced to posthumous punishment after hanging, by having their body's dissected or gibbetted, feared that dismemberment would stymie their soul attaining peace, rest or resurrection in the afterlife.

From a conceptual point of view, the prospect of harming the body after death is a second order symbolic harm to certain spiritual and religious beliefs about a life hereafter (see Chap. 4 and Chap. 5). Furthermore because it is an anticipated symbolic harm that involves imagining what might happen to them in a life hereafter, it also causes psychological suffering.

Posthumous harm becomes posthumous punishment when it deliberately involves state power. Posthumous punishment is a form of retributive justice and involves a further punishment through dismemberment in addition to capital punishment. It is an act of what Bourdieu calls symbolic power/violence (Bourdieu 1992, 1996 cited in Crossley, 2005) where dismemberment through gibbetting or dissection literally involves state authority inscribing a 'mark of infamy' on the criminal corpse.

Symbolic violence also involves a second order intrinsic harm, because the act disremembering the dead causes psychological distress in those that are confronted with that refiguration of memory. The idea of overt intentional symbolic harm to the dead in Chap. 4 constitutes posthumous punishment by the state and needs to be distinguished from posthumous harm, that involves a form of instutional collusion to perpetrate symbolic harm (see Chap. 5).

Fourth Assumption: Posthumous Redemption/Pardoning Is Impossible and Pointless

The corollary of posthumously punishing those shot at dawn was their subsequent posthumous pardoning. This is explored at length in Chap. 4, where the author critically examines the case for and against posthumously pardoning those shot at dawn during the First World War.

Posthumous punishment by being dishourably shot at dawn by one's own men in the First World War was a symbolic harm to soldier's reputation after they are dead. Because capital punishment and its narrative aftermath is a reinterpretation of the narrative events that judges a soldier to be 'deserter' or 'coward' worthy of being shot, it is a secondary symbolic harm. Moreover, those dishonourably executed at dawn in this way, had very real psychological consquences on the men awaiting execution and those relatives that have had to live with this family dishonour.

Posthumous pardoning is an act of symbolic redemption (second order because it is refiguring a narrative of a life lived that has been damaged).

A posthumous pardon is not about rewriting the past as some have argued (Corns and Hughes-Wilson 2002), but a re-evaluation of historic judgments from a new perspective, one that sits alongside a historic judgment. This matters on many levels, not least because it rehabilitates the fidelity of a narrative identity once disremembered. It also acts as a form of forgiveness by taking away the stigma from relatives who have had to live with the dishonour of their ancestor's fate.

SUMMARY

This third chapter builds on the conceptual foundations of the second, conceptualising posthumous harm, punishment and redemption. The major obstacle in theorising the possibility of harm after death is the existence condition. We need to have existed in order to be the subject of harm. It is possible to understand harm in five ways: we can intrinsically harm existing persons; we can harm the transcendent interests of persons that once existed; we can symbolically harm the narrative identity of persons facing their death; we can symbolically harm the memory and biography of the dead; and we intrinsically harm how the dead continue to be remembered through the experience of significant others who feel their dead have been unjustly and harmfully remembered.

If it is possible to conceptualise the possibility of posthumous harm, it is necessary to distinguish posthumous harm from posthumous punishment, where the latter involves a deliberate use of state or symbolic power in order to disremember posthumous identity. Finally, if it is possible to posthumously harm persons through their biography, it is also possible to posthumously redeem their narrative identity through a posthumous pardon.

Discussions in the chapters that follow are marked by a shift in emphasis of concern. Instead of fore-grounding conceptual matters, Chaps. 4 and 5 fore-ground the importance of historical and empirical case studies, occasionally using conceptual insight to further illuminate case study analysis.

References

Belliotti, R. (2013). *Posthumous harm: Why the dead are still vulnerable*. New York: Lexington Books.
Bourdieu, P. (1992). *Language and symbolic power*. Cambridge: Polity.
Bourdieu, P., & Passeron, J. (1996). *Reproduction*. London: Sage.
Corns, C., & Hughes-Wilson, J. (2002). *Blindfold and alone: British military executions in the great war*. London: Cassell.
Crossley, N. (2005). *Key concepts in critical social theory*. London: Sage.
Epicurus. (1940). Letter to Menoeceus. In *The stoic and Epicurean philosophers*. W. J. Oates (Ed.) & Bailey, C. (Trans.). New York: The Modern Library.
Feinberg, J. (1984). *Harm to others*. New York: Oxford University Press.
Feldman, F. (1991). Some puzzles about the evil of death. *Philosophical Review, 100*, 205–227.
Luper-Foy, S. (2002). Death. *Stanford Encyclopaedia of philosophy*. Retrieved January 19, 2006, from http://plato.stanford.edu/enteries/death.
Nagel, T. (1970). Death. *Nous, 4*, 73–80.
NewEpicurean.com—Elemental Edition, *Letter to Herodutus*. Retrieved January 19, 2016, from http://newepicurean.com/epicurus/letter-to-herodotus-elemental-edition/.

Partridge, E. (1981). Posthumous interests and posthumous respect. *Ethics, 91,* 243–264.
Pitcher, G. (1984). The misfortunes of the dead. *American Philosophical Quarterly, 21,* 183–188.
Scarre, G. (2007). *Death.* Stocksfield: Acumen.
Sperling, D. (2008). *Posthumous interests.* Cambridge: Cambridge University Press.
The Murder Act (1751) (25 Geo 2 c 37).
Tomasini, F. (2009). Is posthumous harm possible? Understanding death, harm and grief. *Bioethics, 23*(8), 441–449.
Watson, G. for BBC News. Retrieved January 19, 2016, from http://www.bbc.co.uk/news/uk-england-25841494.

Open Access This chapter is licensed under the terms of the Creative Commons Attribution 4.0 International License (http://creativecommons.org/licenses/by/4.0/), which permits use, sharing, adaptation, distribution and reproduction in any medium or format, as long as you give appropriate credit to the original author(s) and the source, provide a link to the Creative Commons license and indicate if changes were made.

The images or other third party material in this chapter are included in the chapter's Creative Commons license, unless indicated otherwise in a credit line to the material. If material is not included in the chapter's Creative Commons license and your intended use is not permitted by statutory regulation or exceeds the permitted use, you will need to obtain permission directly from the copyright holder.

PART II

Historical Case Studies

'Your legacy is every life you've ever touched.'
  Maya Angelou

CHAPTER 4

Capital Punishment, Posthumous Punishment and Pardon

Abstract Capital punishment is understood in the context of the First World War. Those executed by their own country were shot at dawn under the authority of the British Army Act 1881.
Chapter 4 opens by attempting to understand 'the shot at dawn' policy as deterrent and posthumous punishment. It does so from a multiplicity of perspectives: a fictive account of what being shot at dawn might have been like; a more formal perspective of the facts and reasons behind the 'shot at dawn' policy; its inconvenient longer-term consequences; and examples of a variety of cases where individuals are tried, sentenced and executed. The corollary of posthumous punishment, both short and long term, is posthumous redemption through pardoning. This chapter proceeds by looking at arguments for and against posthumously pardoning those shot at dawn. While both cases have merit, there is a lack of conceptual clarity on how to decide which is the better. With this in mind, there is an attempt to put forward an argument for posthumous pardoning. To end, posthumous punishment and pardoning is understood in its historical long-view, in order to show how such concepts are subject to continuity and change over time.

Keywords Capital and posthumous punishment · Posthumous pardon and symbolic redemption

The Shot at Dawn Policy During the First World War

Execution: The Fictive Reconstruction of Being Shot at Dawn

The condemned private spends his last night in a small room, alone with his thoughts before his execution at dawn. He might be writing painful letters to family and friends. He is also likely to be encouraged to drink heavily in order to be insensible during execution. The private is guarded by two military policemen (MPs or redcaps) and ministered by a chaplain.

The condemned man's commanding officer (CO) orders a company of men to witness the execution, wanting to set an example to other would-be deserters. Meanwhile a firing squad assembles, sick with nerves, in the dawn light. Some of the men know the condemned and have mixed feelings about his fate, some even carrying deep resentment at having to execute him. Their rifles have been pre-loaded—one with a blank—to take some of the individual responsibility away from shooting their fighting pal.

The condemned is led, blind drunk, to a post by two redcaps, his hands tied behind his back. The lieutenant waits at the side of the shooting party, with a medical officer (MO). The lieutenant (Lt.) gives the order to shoot the prisoner. Some deliberately shoot wide. Two of the men vomit on the spot. The MO checks the prisoner over and concludes that the private is mortally wounded but not dead. The young lieutenant, with shaky hands, administers the *coup de grâce*: a bullet to the head.

A military ambulance stands by to take the corpse off to be buried. That same evening the battalion colonel writes a letter to the private's parents informing them that their son has been shot at the front. He leaves the message deliberately ambiguous, sparing the man's family any difficult feelings about his execution (see also Johnson 2015).

Punishment and Execution in Historical Context

The corollary of being a 'good soldier', a disciplined effective fighter that followed orders, was a 'bad soldier' that threatened discipline and the army's effectiveness as a fighting force. Being a 'bad' soldier inevitably led to punishment of various forms: field punishments, and a court martial for offences punishable by death.

In practical terms, military law was enshrined in the Manual of Military Law (MML), which was 'a formidable weighty tome', providing commentary on how the law stood in 1914, originally articulated in the statutory Army Act 1881. Every possible offence and misdemeanour, large and small, and how it ought to be interpreted, was set out in the MML (Corns and Hughes-Wilson 2002, pp. 44–46). While the MML was not something that soldiers, or indeed most officers, carried around with them, soldiers were made aware which offences were punishable and for what reasons: such information being available to them in summary form, at the back *the soldier's small book*—which they received on enlistment (Corns and Hughes-Wilson 2002, p. 342).

There were four kinds of court martial in 1914: the Regimental Court Martial (RCM) ; the District Court Martial (DCM); the General Court Martial (GCM); and, the Field General Court Martial (FGCM) (Corns and Hughes-Wilson 2002, p. 89). Most of those sentenced to be executed and shot at dawn were tried and sentenced by the FGCM, as this was the swift arbiter of military justice in the field and designed to expedite serious crimes normally reserved in peace time for the DCM or GCM, which was presided over by a legally trained Judge Advocate (Corns and Hughes-Wilson 2002, pp. 93–94).

The FGCM would consist of three officers: normally a brigade commander (a major or above), a captain and a lieutenant. While there were usually no professional lawyers on an FGCM, the case was reviewed in a post-sentencing procedure by a Deputy Judge Advocate General (DJAG), who made sure there were no miscarriages of justice on technical legal terms (Putowski and Sykes 2014, p. 17). The accused could choose a 'prisoner's friend', usually an officer, to help defend him (Putowski and Sykes 2014, p. 14). This practice became more commonplace as the war went on. Even so, over 10% of those executed did not have representation, known euphemisically as a prisoner's friend (Putowski and Sykes 2014, p. 20).

The prisoner could object to the composition of the sitting panel if he thought an officer might prejudice his case. The officer who convened the court martial could not sit on it and the most junior officer was the one who voted first, ensuring that he would not defer to his superiors judgment (Holmes 2004).

After sentencing by Court Martial (CM) there was important post-sentencing procedure where the accused had to wait in order for officers higher up the chain of command to append comments on the sentence of the CM, before it was passed to the DJAG and Commander-in-Chief

(C-in-C). Their comments held weight and could influence the C-in-C and his final decision on whether or not a guilty verdict was to lead to the death sentence.

This system was introduced post-April 1915 and included the following categories of appended comments:

- The medical record of the soldier. A judgement and comment from an MO on the fitness of the soldier for action;
- Whether or not the accused was a good soldier. Was he a good fighting man with a good character? The soldier's CO contributed to this category;
- The state of regimental discipline. Was the verdict to be upheld because the man needed to be made an example of for the sake of discipline? Such comments were often reserved for senior officers of the rank of Brigadier General or above (see Putowski and Sykes 2014, p. 18).

An FGCM could only recommend a particular verdict. Over and above the judicial decision the execution of sentencing was still subject to post-sentencing procedure.

The first category was subject to errors of medical judgment, where those regarded as cowards may have actually been affected by shell-shock. However, those we now consider wrongly judged as cowards could well have been 'correctly' diagnosed as not suffering from shell-shock—given the historic limits of medical understanding at the time. We need to be careful not to re-write history by overriding technically correct medical decisions then, with new medical understanding about shell-shock and post-traumatic stress that we have today.

The second category was probably more problematic in respect to ensuring fairness. Judgments about what constituted a 'good' as opposed to a 'bad' fighting soldier were often based on hearsay evidence pased up the line of command and that often reflected a set of values held by a senior officer in the post-sentencing process about how the rank and file were expected to behave. Senior officers' 'damning remarks' could have had an influence on the C-in-C who was the final judge as to whether a soldier was to be executed. Comments from the 'château generals' were filtered through a very different cultural and class experience and could be particularly harsh when cowardice was suspected (Putowski and Sykes 2014).

By far the most unfair criterion was the third, where some decisions to execute came down to whether or not those convicted should be made an example of for the sake of others. Those tried and convicted under military law were all subject to fortune and whether officers higher up the chain of command deemed it necessary to execute for the sake of example and regimental discipline. Regimental discipline seems to have been connected to how much pressure the Army was under, so at the opening of the Somme campaign (July 1916) Sir Douglas Haig's judgment to execute has in hindsight been criticised as 'defective' (Putowski and Sykes 2014, p. 12). In sum, the final decision to execute or commute might be considered fairly arbitrary in some cases—just deserts relying on what was militarily expedient on top of what was technically correct in terms of military justice served.

Despite such concerns about post-sentencing, 9 out of 10 sentenced of a capital crime had their sentence commuted by the C-in-C who had the final say in capital cases (Putowski and Sykes 2014; Corns and Hughes-Wilson 2002).

Corns and Hughes-Wilson break down that statistic as follows:

> During 4 August 1914 to October 1918 there were approximately 238,000 courts martial resulting in 3080 death sentences. Of these only 346 were carried out, which break into the following categories of offences on active service:
> Mutiny 3
> Desertion 266
> Cowardice 18
> Disobedience of a lawful order 5
> Sleeping at post 2
> Striking a superior officer 6
> Casting away arms 2
> Quitting post 7
> Murder 37 . (Corns and Hughes-Wilson 2002, pp. 103–104)

Aftermath of the Shot at Dawn Policy—Some Critical Reflections

It is deeply questionable whether or not the army's short-term deterrent policy was successful.

Executing soldiers for a range of military offences was met with mixed reactions, largely depending on how the rank and file perceived the fairness of justice done. On more than one occasion, soldiers who had served with those sentenced harboured anger and resentment against their commanders for making such decisions, especially where there was a spike of executions within a battalion for the sake of example.

The Australian army didn't have a capital punishment policy (Corns and Hughes-Wilson 2002, p. 391) and seemed by comparison to operate well without it. Another comparison is with the German army. The German army did have a court martial system and capital punishment. Unlike the British and some Commonwealth nations, it involved a much more rigorous court martial system, where legal professionals for defence and prosecution maintained very high standards of military justice in front of a legally trained judge (Barton 2016).

One of the most concerning aspects of British military justice during the First World War was that it did not employ legal professionals, given the fact that there were, potentially at least, plenty available (Corns and Hughes-Wilson 2002, p. 94). CMs were rarer in the German army, there being a policy of officers disciplining their men in the field as they saw fit (Barton 2016). Moreover, compared to the British army with 346 executed, the German army executed only 18 men (Barton 2016; Corns and Hughes-Wilson 2002).

Psychologically it is not at all clear that the deterrent effect of execution motivated soldiers. As Corns and Hughes-Wilson have argued, it is not fear of being shot by one's own side that motivates soldiers to fight, but a discipline that comes from working in a 'tightly knit group and fighting for each other in a life threatening crisis' (Corns and Hughes-Wilson 2002, p. 456). Finally, looking back, it seems deeply illogical for an army to pride itself on training its soldiers only to undermine this confidence by anticipating that they should be made an example of for ill-discipline during war.

Probably the saddest consequence of the policy flowed from the hardship and shame that this policy caused on the family left behind. The widows of those shot at dawn were often denied an army pension. Relatives of the executed were left stigmatised, as the stain of familial dishonour remained long after the *raison d'être* of deterrence for execution had expired.

The intention behind the policy was not to punish or dishonour the family of those executed. Commanding officers often went out of their way to avoid telling the next of kin the exact circumstances surrounding the death of a loved one. There were a few good reasons why this was done. The most pragmatic was that it was bad publicity for the army;

being censorious about the execution policy would have undermined the army's recruitment policy, which relied on Kitchener's volunteers once trained professionals—the British Expeditionary Force (BEF)—required significant reinforcement as the war progressed.

The deterrent motivation behind these military executions was short term. That is, the posthumous punishment of being dishonoured for being shot as a 'coward' or a 'deserter' was designed to instil order and discipline in the fighting troops.

The character of such retributive justice appears to have no long-term malicious intention. There is ample evidence to show that COs were often compassionate to families by sparing them the truth, that post-war governments were sympathetic, if secretive, about the details of those executed, and that the War Graves Commission made no distinction between those fallen in action and those shot at dawn.

The secretiveness around the shot at dawn policy, however, caused problems after the war. It fuelled parliamentary scrutiny and eventually galvanised campaigners years later into uncovering the truth and demanding justice (Corns and Hughes-Wilson 2002, pp. 403–447). It also angered relatives who felt deceived as to the true fate of a loved one who had been shot. George Ingham's father, for example, felt angry at being deceived as to why his son had been shot. When the War Graves Commission re-contacted George's father and asked him what he would like carved on his son's headstone, the angry parent who had learned truth about his son, requested a stand out inscription that reads: '*Shot at dawn one of the first to enlist a worthy son of his father*' (Corns and Hughes-Wilson 2002, p. 259).

It is worth distinguishing intended and unintended outcomes. The intended military outcome of the shot at dawn policy was to execute according to the rule of military law and dishonour for the sake of example. The unintended outcome of the shot at dawn policy was that this human stain of dishonour lingered for years afterwards, affecting families for many generations. For example, Terence Highgate the great nephew of Thomas Highgate who was shot for desertion on 8 September 1914, was still preoccupied, in 2014, with having his relative's name inscribed on the local war memorial in Shoreham, Kent (Kentlive 2010).

Many years after the Great War had ended, governments of different stripes and colours have wrestled with pardoning those executed, refusing, whether sympathetic to their plight or not, to grant a posthumous pardon for technical legal reasons. In the 1980s, campaign pressure mounted as John Hipkin, a retired teacher, formed 'the shot at dawn campaign' supported by relatives of those executed. Finally in 2006 the Labour Defence Secretary of State, Des Browne, offered a blanket pardon for 306 shot at dawn (excluding 40 of 346 who had been shot for murder and mutiny).

Retributive Justice: Some Individual Case Studies

Harry Farr: Shot for Cowardice

The Harry Farr case is summarised from two secondary sources: Corns and Hughes-Wilson (2002, pp. 202–205) and Putowski and Sykes (2014, p. 121).

On 7 May 1915 Farr's battalion received information of a possible attack. All men were to be issued gas masks. On 8 May the battalion took up positions in the assembly trenches and preparatory bombardment took place at dawn on 9 May before the planned assault on Auber's Ridge. Farr's regiment, the 2nd Yorkshires, became pinned down by continual shellfire in the assembly trenches. This also happened the following day, and on 10 May the assault was cancelled. Harry Farr was evacuated to Boulogne on 11 May where his nerves were reportedly so badly affected that he could not hold a pen. After convalescence he was sent to the 1st battalion of the 1st West Yorkshires. Farr's battalion was not involved in the initial phases of the battle of the Somme in July 1916. However, in August they moved to advanced positions, and on 16 September took over the front line near the Quadrilateral. On 18 Sept they attacked the Quadrilateral and took 151 casualties. Just before the attack Harry Farr's nerves failed.

Harry Farr had no 'prisoner's friend' (defence representative) at his trial on 2 October. The evidence presented against him was largely uncontested by Farr, although he did add some authentic colour to one of the key witnesses against him at the FGCM, Regimental Sargent Major (RSM) Haking.

Haking first saw Farr at the transport lines at 9 a.m. on 17 September 1916 after falling out the previous night due to reported sickness. Farr refused treatment because he was not wounded. Haking then ordered Farr to the front. Farr did not go and by 11 p.m. that evening when he still hadn't gone Haking tried to compel Farr. Haking told Farr he would be shot if he did go to the front. Farr in reply said, amongst other things, 'I cannot stand it' and 'I am not fit to go to the trenches' and when compelled to see the MO under escort said: 'I will not go any further that way'. According to Farr his refusal to go to the front incensed the RSM so much he said 'You are a fucking coward and you will go to the trenches—I give fuck all for my life and I give fuck all for yours and I'll get you fucking well shot' (cited in Corns and Hughes-Wilson, p. 203). Farr added that he would have co-operated if his escort had not pushed him.

The court asked Farr whether he had taken the opportunity of reporting sick between 16 September and 2 October. He replied that whilst he had had the opportunity he had not reported sick, because 'being away from the shell fire I felt better'. In his defence Acting Sergeant Andrews said Farr had reported sick with his nerves on April 1916 when the MO had kept him at the dressing station for two weeks. This happened again on 22 July, when he had been discharged the following day. Crucially, the doctor who had treated Farr then was unable to give evidence as he had been wounded.

The court found Private (Pte) Harry Farr guilty as charged. He was sentenced by the FGCM to be shot. Before the sentence was carried out commutation was possible—the case for mitigation going up the line of command. His company commander wrote 'I cannot say what destroyed the man's nerves but he has proved on many occasions incapable of keeping his head in action and likely to cause panic. Apart from his behaviour under fire his conduct and character are very good' (cited in Corns and Hughes-Wilson, p. 204).

Brigade, division and corps commanders all recommended that sentence be carried out; Lieutenant General (Lt Gen.) Lord Cavan, commanding XIV Corps, fatefully added 'The General Officer Commanding 6 Division informs me that the men know that the man is no good' (cited in Corns and Hughes-Wilson, p. 204). In some post-trial correspondence, Dr Capt. Williams of the Royal Medical Army Corps (RMAC) wrote to the adjutant of the battalion saying that Farr's 'mental and physical conditions were satisfactory'.

Harry Farr was shot for cowardice and executed at Carnoy on 18 October 1916. He refused to be blindfolded and looked the firing squad in the eye. The signature of Capt. A. Anderson (RMAC) who witnessed the execution was shaky, suggesting perhaps that that the doctor was shocked by what he witnessed.

What should we make of the Harry Farr case?

Farr's refusal to fight could have warranted a judgment of cowardice for a number of historical reasons. It was much easier to judge someone to be a coward than it was to prove otherwise. The fact that shell-shock was not properly distinguished from cowardice in the Farr case, rested on the fact that there was no compelling medical evidence that Farr was shell-shocked *when* his nerves gave way on the fateful night of 17 September 1916. One of the tragedies of the Farr case is that he wasn't seen by an MO directly after he had reported sick after his nerves failed on 17/18 September. His later diagnosis by an MO, two weeks after he had originally reported 'sick', failed to pick up on his condition.

Without corroborating medical evidence pertinent to his particular offence, Farr's self-assessment that he was 'unfit to fight' was insufficient to save him from execution. Historical anecdotal evidence of shell-shock as mitigation did not prove he was not shell-shocked on 17/18 September.

Harry Farr, in the post-sentencing procedure, was unlucky. Lt Gen. Lord Cavan comment that he had heard the man was 'no good' would have counted as influential and a potentially damning remark.

Given the standards of military justice meted out at the time, the case is legally sound and unremarkable. Even though Lord Cavan's remark post-sentencing is harsh, and could easily be misinterpreted, it was accurate; Farr probably was 'no good' as in not fit as a fighting soldier.

The disturbing thing about the Farr case was that there was no benefit of doubt given to the ample evidence that he was affected by shell-shock, at least historically. Farr and those who spoke up in his defence provided sufficient circumstantial evidence that Farr was likely to have had shell-shock and that the MO's testimony, taken two weeks after his nerves failed at the front, was therefore not wholly reliable. However, given all the aforementioned circumstances that played against him, it is not altogether surprising that the FGCM ruled against him and that post-sentencing delivered no commutation in the sentence.

From a more ideal perspective, where our sense of natural justice is less encumbered by historical context, the case is worrying for a number of reasons.

Farr's historic medical record should have given the court ample reason to doubt the perfunctory assessment of an MO who saw Farr two weeks after his nerves gave way.

Farr was one of the 10% who did not have a prisoner's friend to help defend him. He was subject, like others before and after him, to a highly dubious post-sentencing procedure that circumvents the proper appeal process that one expects from civilian law.

Lord Cavan's comment about Farr is no more than military gossip.

Finally, Farr's case needs to be understood in a strategic military context: that is, Farr's case occurred during the long Somme offensive where army commanders would have been less inclined to show mercy and more likely to recommend execution in order to set an example. This arguably taints justice and succumbs decision-making to military fortune and expediency.

Ingham and Longshaw: 'Pals' Shot for Desertion
The Ingham and Longshaw case is also summarised from two secondary sources: Corns and Hughes-Wilson (2002, pp. 257–259) and Putowski and Sykes (2014, pp. 138–140).

Pte Albert Ingham and Pte Alfred Longshaw worked together before the war for the Lancashire and Yorkshire Railways. After enlisting as serving privates in No.11 platoon, 'C' Company, the 18 Manchester's, they travelled via Egypt to the Western Front on 18 November 1915. After experiencing active service on the Somme, the two 'pals' were posted to a brigade machine gun company. In the second week of October the two men, after hearing that they were being prepared for a move to the trenches, decided to abscond. Avoiding the authorities they managed to stow away on a Swedish ship at Dieppe. The two pals were caught when Longshaw unsuccessfully tried to pretend that he was an American travelling to Spain.

Both Ingham and Longshaw were convicted and sentenced to death. In his defence Ingham said:

> I was worried at the time about the loss of my chums also about my mother, being upset, through hearing bad news of two of my comrades. I plead for leniency on the account of my service in France for 12 months and my previous good conduct. I beg for the chance to make atonement. I left with my chum firstly to see those at home and then to try and get into the Navy along with his other brother, who is serving there. (cited in Corns and Hughes-Wilson 2002, pp. 257–258)

The evidence against them was overwhelming and Ingham's plea for atonement was ignored. During the post-sentencing procedure the commanders up the chain of command all supported the death sentence, Brigadier general Lloyd, commenting: 'A well-thought out plan of escape from service is disclosed and a man who commits such a crime deserves the extreme penalty' (cited in Corns and Hughes-Wilson, p. 258). Ingham and Longshaw were shot on 1 December 1917.

What should we make of the Ingham and Longshaw case?

To begin it is worth contrasting a difficult case of cowardice from a more straightforward case of desertion. Whether a man is a coward or not involves making some kind of subjective judgment about their internal mental state. This is difficult and risks error. On the other hand, making a judgment about whether a soldier is deserter or not is, by comparison, fairly simple because it relies on a series of objective and straightforward observations. Namely, is the suspected deserter at the front with his company where he is supposed to be? Is the said deserter disguised in civilian clothes?

Given these simple criteria, Ingham and Longshaw were obviously guilty as charged; they were on a boat heading out of Dieppe and they were well disguised in civilian clothes. Moreover, their elaborate disguise and mode of transport all pointed to what Brigadier Lloyd described as a 'well-thought out plan' of escape.

What is the case for mitigation?

From the perspective of military justice at the time, the case for mitigation is weak. Unlike the Farr case, where there was a significant doubt as to whether he was indeed a coward, Ingham and Longshaw were definitely deserters with a plan of escape. According to historical standards, justice was served.

However, again with benefit of hindsight, and armed with the zeal of natural justice and retrospective fairness, a case for mitigation is possible. Like Harry Farr, Ingham and Longshaw were not adequately represented with a prisoner's friend at trial. Ingham's plea for mercy and atonement, whilst not successful, seemed plaintively honest and reasonable. He was a first-time deserter with a good record who had temporarily lost his fighting spirit because he was grieving for his 'pals'. Furthermore, when compared to serial deserters who were often given lesser (physical) field punishments, court martialling and executing him and his pal seems logically inconsistent and unfair given that they were first-time offenders with decent military records.

Rogues and Murderers

The most difficult cases are rogues who had no intention of following the rules and those who committed either murderous or treasonable acts. Again this historical material is summarised from Putowski and Sykes (2014) and Corns and Hughes-Wilson (2002).

Take the example of 'the rogue' Corporal (Cpl) George Latham. Cpl Latham arrived in France on 22 August 1914, and in the confusion of the retreat of Mons deserted. He deserted after three days on active service.

From the very start of his arrival in France, Cpl Latham showed little spirit to fight. He covered his tracks and quite happily lived behind the front line in Nieppe with two women who lived on the same street, rue d'Armentières. His previous paramour, Mme Chilbrae, probably out of jealousy for transferring his affection to Mme Cambiers, notified the MPs of his secretive life by showing them his mail sent to her address. Latham presented a convoluted and incredible defence that was not believed at his court martial. On 22 January 1915 he became the first Non-Commissioned Officer (NCO) to be executed for long-term desertion from the BEF (Corns and Hughes-Wilson 2002, p. 121).

What should we make of the Cpl Latham case?

Latham was clearly a long-term deserter, who unlike Ingham and Longshaw had no intention whatsoever to fight for King and Country. He had no fighting record in the First World War and lived a salacious lifestyle, while his comrades at arms risked their lives in the trenches. From a military point of view Latham was a 'rogue', and a bad example to others, showing no desire to return to fight. There was a clear case for him to be made an example of.

With the benefit of hindsight it is understandable that someone might run and hide rather than risk a very good chance of getting killed in battle. However, this is not a strong case for mitigation, given the historical context of what was expected of soldiers at the front, and how such behaviour is likely to have been viewed as its likely impact on others.

The case for mitigation is even more difficult for murderers who were shot at dawn. Clearly murder is not only a risk to military order and discipline, it was also a capital punishment offence in the civilian courts at the time. It is hard, therefore, to argue that justice—from a historical perspective at least—was not served in these cases.

With the benefit of hindsight some of these individuals appear to have been mentally damaged by the brutal war in which they fought. Therefore they should not have been subject to the same treatment as 'sane' persons who committed murder (see also Putowski and Sykes 2014).

Critical Reflection on Posthumously Pardoning Those Shot at Dawn

The Historical Case for a Posthumous Pardon: The Putowski and Sykes Thesis

Putowski and Sykes (2014) in *Shot at Dawn: executions in world war one by authority of the British Army Act* believe that the execution of soldiers during the Great War was wrong. In the foreword to this work they state:

> After very careful consideration, a decision was reached (just prior to publication) to press for the complete exoneration of these men. Convictions should be quashed and the men pardoned. (Putowski and Sykes 2014 p. 9)

In the all-too-brief introduction, Putowski and Sykes provide a deeply critical account of justice served. Much of Putowski and Sykes's criticism of the justice these men received is based on the 'unforgiving military judicial system'.

One of their main complaints concerns the class bias in which justice was apportioned. They imply that the system was deeply unfair: officers overwhelmingly drawn from the upper classes dispensing capital punishment on privates and NCOs, predominantly drawn from the working class (Putowski and Sykes 2014, p. 16). This is certainly borne out in the statistics—the vast majority executed were ordinary soldiers and NCOs: 3 officers were shot at dawn out of a total of 306 ordinary soldiers and NCOs that were eventually pardoned in 2006.

The other major criticisms that Putowski and Sykes dwell on is the unfairness of the military justice system. In their own words:

Records relating to the number of these cases highlight just how arbitrary the decision to confirm the death sentence could be. The inference drawn from these records is that after the opening of the Somme in July 1916, the judgment of Sir Douglas Haig was defective when he decided to have certain men executed... (Putowski and Sykes 2014, p. 12)

This apparent unfairness, they argue, was exacerbated by the lack of proper representation for those accused of a capital crime during the First World War. They are among the first to point out that 10% of those accused of a capital offence did not have representation (a prisoner's friend). Lack of representation was a serious obstacle in conducting an adequate defence. This was aggravated by defendants' representatives not cross-examining the prosecution witness, which according to Putowski and Sykes was usually the case.

Furthermore, Putowski and Sykes point out the probable unfairness of pitting ill-educated soldiers against their superiors in their respective roles as prosecutor and judge. As Putowski and Sykes (2014) put it 'speaking up to hostile authority for a soldier conditioned to obey its judgment must have been an awesome ordeal, whilst remaining silent could easily be mistaken for an admission of guilt' (Putowski and Sykes 2014, p. 15).

Putowski and Sykes are also very critical of the post-sentencing procedure (Putowski and Sykes 2014, pp. 11–23). Post-sentencing is dependent on extra-judicial decisions made by senior officers that are not necessarily pertinent to the circumstances in which justice should be served. This makes sentencing into something of a lottery as already argued, in which 'bad' soldiers in military terms are condemned when an example is needed for the sake of regimental discipline. The problem was further exacerbated by the so called 'château generals', who had little knowledge of the particulars of the case and had little empathy for the plight of ordinary soldier on the front line.

The Historical Case Against a Posthumous Pardon: The Corns and Hughes-Wilson Thesis

The case against posthumous pardoning is articulated by Corns and Hughes-Wilson in *Blindfold and Alone*. On more conceptual grounds they hold the opinion:

With our more modern and enlightened views it is hard not to feel compassion for those who died as a result of such laws. The fact that many today feel *now* that it is wrong, however, does not make it wrong by the standards of its day, any more than we can deplore the inability of doctors in the Great War to carry out blood transfusions or their attempts made in good faith, to solve psychiatric trauma by applying electric shocks which amounted to sheer torture. We cannot undo the deeds of the past. (Corns and Hughes-Wilson 2002, p. 457)

Cathryn Corn's co-author Colonel John Hughes-Wilson goes further, explicitly warning against the dangers of re-writing history:

If these men were alive today, we would not kill them. But we must be very wary about applying our modern sentiments and values to the 1914-18 war. We cannot re-invent the past to suit ourselves today. And even now we expect our servicemen, and women, to do what they presumably signed up to – risk their lives and fight. (Corns cited in Peter Taylor-Whiffen, 2011)

The more specific empirically orientated arguments against posthumous pardoning revolve around the paucity of material available on which historic judgments are made, and the sheer variety of offences where one-size-fits-all justice is totally inappropriate.

On insufficiency of evidence Corns and Hughes-Wilson argue:

In the first place, by definition any pardon or quashing of sentences demanded by the pardons campaigners must be a selective judgment. Given political will, there would be no insuperable obstacle were there to be sufficient evidence on which to make a judgment. However, the files today are not merely incomplete: only those relating to the men who were actually executed survive … we have no way to compare the records of those confirmed for execution and those reprieved… As successive governments have discovered, selective judgments would risk being mere arbitrary opinions. (Corns and Hughes-Wilson 2002, pp. 457–458)

What Is a Posthumous Pardon for?

The key problem in deciding on a position for or against posthumous pardoning is conceptual as well as empirical.

The historians discussed tend to draw upon strong views without fully scrutinising their assumptions about what a posthumous pardon is for. This has left the discussion somewhat polarised.

In order to rectify this lacuna in the debate it is helpful to explore some conceptual assumptions on which posthumous pardoning rests, as well as mobilising empirical arguments for why posthumous pardons may or may not be justified.

A strong opinion against the granting of a blanket pardon – by the then Defence Secretary Des Browne—is articulated by the military historian Correlli Barnett. In his words:

> "These decisions were taken in the heat of war when the commanders' primary duty was to keep the Army together and to keep it fighting. They were therefore decisions taken from a different moral perspective... For the people of this generation to come along and second-guess decisions taken is wrong. It was done in a particular historical setting and in a particular moral and social climate. It's pointless to give these pardons. What's the use of a posthumous pardon?" (Cited in Fenton 2006)

While Barnett is right to suggest that decisions to execute those shot at dawn need to be understood in historical context, from within 'a very particular moral and social climate', he is mistaken to argue that this necessarily leads to 'second guessing' historic decisions. This would be true if posthumous pardoning necessarily involved re-writing history. However, it is also possible to re-evaluate historical decisions in the present through the mechanism of a posthumous pardon—a view that Barnett does not consider.

Those of us who support posthumous pardoning are not necessarily re-writing history to suit the moral standards of the present. Understanding historic justice does not preclude re-evaluating its moral force now. The normative force of the past is not hermetically sealed; its effect has an influence on present generations who have to live with decisions that condemned their ancestors. This is particularly difficult when normative historic decisions no longer stand the test of time. Whilst it is, of course, important not to rewrite the past to suit the present, it is perfectly acceptable to re-evaluate its normative influence, especially when such influence shames contemporaries still affected by it.

Corelli's comment that a 'posthumous pardon is pointless' is superficially true only. It is pointless, presumably, because it has already happened? His more serious criticism concerns re-writing the past. Again, this is true if this is the only way of conceiving of what a posthumous pardon is for. If so, this begs the question: is there really anyone who might actually think rewriting history is a good idea? This feels very much like a 'straw man' argument (setting up and refuting an opponent's argument that is not actually advanced by that opponent).

This author believes that posthumous pardoning is both conceptually intelligible and plausible in certain specific circumstances. They matter for at least four reasons.

1. *A posthumous pardon restores the symbolic narrative fidelity of a life remembered. This is particularly important in the rehabilitation of the narrative identity of persons who once existed.*

Why is this important? Take the example of Harry Farr again. Farr's execution for being a coward is likely to have been a mistaken judgment. Again, we do not have to rewrite history in order to appreciate this.

We know today that shell-shock should under no circumstances be conflated with cowardice. Shell-shock is a failure to cope under fire and manifests as post-traumatic stress disorder (PSTD) and battle fatigue. It is a collapse of the will and an inability to cope in the stress of battle. This is quite different from cowardice, which is about fearfulness and faintheartedness, where one is unwilling to fight.

Rewriting the past in the present is problematic because it destroys it. Re-writing the past in this sense is about over-writing what we think should have happened given the values and understanding we have today. This is not appropriate if one is to understand Farr's case historically. Historically speaking Farr's trail, sentence and execution was just, if somewhat harsh, given the historical context of judicial decision making and medical knowledge available at the time. However, while as a historian one might accept this view, it should not preclude the fact that it is perfectly legitimate to revaluate certain 'facts' about the effects of shell-shock, that historical actors *could* not have fully grasped then.

Re-evaluating the past in the present is not the same as rewriting history. The truth of Harry's shell-shock is subject to forensic re-evaluation as our understanding of certain conditions like shell-shock has evolved over time. It would be perverse and unfair to the relatives of Harry Farr not to acknowledge the re-evaluation of Harry's reputation, as our understanding of his likely condition then has grown now. Simply put, we can understand and respect historic decisions made then, without the burden of being morally bound by them today when deliberately re-evaluating them.

Note re-evaluation does not involve condemning those who judged, convicted and executed Farr, given a deep appreciation of historical context and its interpretive limits. However, *re*-membering the past is not some scholarly 'God trick' (Harraway 1988 cited in Harding, 2004) by which the past is forever preserved in contextual aspic, where no one can re-evaluate past decisions that seem blatantly flawed today. History as a 'dialogue between past and present' (Carr 1990) is not only scholarly insight—where there is a desire to provide fine reinterpretations in order to understand historical contexts in their own right—but also a moral one—where there is sometimes a need to normatively re-evaluate decisions made at that time that we can no longer hold or justify today.

So, what is the point of morally re-evaluating the past in the present? It is useful to gain a comparative perspective, where the rehabilitation of Farr's reputation, for example, is a gesture to those still living today who have to live with past decisions that do not stand the test of time and are now considered unjust.

We need to distinguish *historic justice*—in which social actors did their best to reach procedurally correct judgments based on a limited understanding of facts within definite historical constraints—from *natural justice*—where we do have the benefit of hindsight to re-evaluate the inherent fairness of historic cases.

2. *Posthumous pardoning is a form of symbolic redemption from the unintended consequences of posthumous punishment.*

From the perspective of historic justice the military executions were intelligible and generally consistent with historic values and military justice and procedures at the time. For this reason it is not clear that Putowski and Sykes have a strong case to quash these convictions.

However, the unintended consequences of the army's 'shot at dawn policy' went well beyond its intended effect. What was not intended was that soldiers rightly convicted by the court, and who had their death sentence upheld by the C-in-C as an example to other fighting soldiers, would be dishonoured for years to come. This was further compounded by their living relatives that carried the historic shame of the original dishonour.

Posthumous pardoning is a form of symbolic redemption; a way of redeeming a dishonour that is carried over from one generation to another.

3. *Posthumous pardoning has an important role in forgiveness and reconciliation. It forgives historical actors, as well as those emotionally affected by its decisions.*

Today, the shot at dawn campaign has been re-evaluated as a symbolic harm to the memory of the dead. This is complicated, not least because we should not conflate what was considered rightly just then, with what might no longer be considered just today.

The most compelling reason to re-evaluate the past in the present is when it causes unnecessary and unjust suffering today. The symbolic and narrative redemption of reflexive biography has a way of conserving the fidelity of memory as well as having the redemptive effect of removing the human stain of inherited shame.

For example, Harry Farr's widow, Gertrude, struggled to cope after his death with a young daughter and no army pension. Gertrude and his daughter (also called Gertrude) unsuccessfully tried to clear Harry Farr's reputation through the courts. The psychological burden of this failure was an intrinsic harm—albeit of a secondary order—to the relatives of the dead. Des Browne's blanket pardon—which included Harry Farr—could not have come any sooner for Harry Farr's daughter Gertrude, who at the age of 93 lived to see her father forgiven. In her own words: 'I am so relieved that this ordeal is over and that *I can be content* knowing that my father's memory is intact' (cited in Norton-Taylor 2006 [Author's Italics]).

4. *It is a way of society understanding itself in relation to its past. By re-evaluating the past in the present we are acknowledging that what was considered wrong then is not something that we can view with the same disdain today. It signals to the world that while we understand why certain individuals were treated in the way they were, we can no longer view these people in the same way given what we know today.*

To really appreciate this perspective we need to think beyond the rather crude dichotomous reasoning which many historical commentators around this subject fall prey to when articulating a case for and against posthumous pardoning.

On the one hand there are those who deplore its injustice and want to rewrite past justice with the benefit of hindsight (Putowski and Sykes 2014) and those who want to preserve, understand and legitimise historic judgments (Corns and Hughes-Wilson 2002).

Putowski and Sykes's revisionist stance appeals to our sense of natural justice, according to which decisions made then seem deeply unfair according to the standards of justice that we might expect today. To reiterate, the authors call for these verdicts to be quashed and overturned. This forces them into making the case that there were miscarriages of historic justice. Unfortunately, while one can have much sympathy for their clarion call for natural justice, their historical case for a miscarriage of justice is perfunctory and inadequate. In the end, Putowski and Sykes neither provide the new evidence for over turning historic cases, nor provide a forceful enough argument from natural justice.

Corn's and Hughes-Wilson's post-revisionist thesis appeals to the more diligent historian who seeks to understand decisions from within the historical context that conditioned them. However, their argument that posthumous pardoning per se is a form of re-writing history is only true if that exhausts all the options of what we think a posthumous pardon may be for. As argued earlier, there is another way of thinking about posthumous pardoning that Corn's and Hughes-Wilson do not properly consider; re-evaluating memory is a deeply normative process, one that needs to be distinguished from the scholarly pursuit of understanding historical context for its own sake. Where as the former involves re-evaluating how we now stand in relation to our past, the latter involves remembering the historical context and limitations to decision-making. Re-evaluation is not the same as re-writing history because one wishes it were different.

The better argument against posthumous pardoning that Corns and Hughes-Wilson put forward is a practical one. One problem they identify is the sheer complexity and variety of different offences for which court-martialling operated, so that just one kind of retrospective re-evaluation of justice does not fit all cases (Corns and Hughes-Wilson 2002, pp. 446–447). This definitely provides a significant hurdle for the design of an appropriate mechanism.

Des Browne's solution was to design a blanket statutory pardon for the 306 shot at dawn with certain important caveats that had inbuilt checks and balances.

In Des Browne's own words:

> In each case, the effect of the pardon will be to recognise that execution was not a fate that the individual deserved but resulted from the particular discipline and penalties considered necessary at the time for the successful prosecution of the war. We intend that the amendment should as far as possible remove the particular dishonour that execution brought to the individuals and their families. However, the pardon should not be seen as casting doubt on either the procedures or processes of the time or judgment of those who took these very difficult decisions. (Hansard, Sept. 2006, HC, co. 135WS)

Des Browne tried to find a reasonable common ground that recognised why most of those shot at dawn could be pardoned, without granting *all* of those shot at dawn a pardon. Browne's statutory pardon deliberately excluded the most difficult cases, such as murderers and mutineers. Moreover, he included a caveat that recognised the importance of historic justice, which deliberately did not bring into doubt the standards, procedures and judgments that condemned these men in the first place. In other words, it could be argued that Browne's blanket pardon cleverly sidesteps the difficult legal ground, focussing more on the moral case that re-evaluates 'just deserts' in terms of how we understand the stresses of the First World War today. In doing so, Browne emphasised the importance of posthumous pardoning in the lives of the families who continue to suffer because of what happened to their relatives. Again, in Browne's own words:

> Although this is a historical matter, I am conscious of how the families of these men feel today. They had to endure the stigma for decades. This makes this a *moral* issue to, and having reviewed it, I believe it is appropriate to seek a statutory pardon. (Cited in Fenton 2006)

The most disappointing aspect of Corns and Hughes-Wilson's thesis is that they seem to be completely blind to the moral case for posthumous pardoning and the vital role this plays in lives of the surviving relatives who have carried the burdensome pain of their ancestors' dishonour.

There are two more general arguments against posthumous pardoning. While both arguments do not invoke this particular case study, they are, nonetheless, conceptually relevant to understanding it.

The first argument against posthumous pardoning is put forward by Matthew Parris in *The Spectator* (2014). Parris is of the opinion that posthumous pardoning is subject to what is called a 'slippery slope argument'. That is, allowing one seemingly harmless pardon, might lead to an avalanche of currently unthinkable cases becoming accepted today. While Parris is right to point out the danger in the case of Alan Turing, it is easily countered by putting some robust conditional criteria, distinguishing historically specious claims from morally 'live' cases that still matter to those living today. Not all historical cases for posthumous pardoning are morally relevant today, which cuts down on the potential avalanche of cases. As Browne makes perfectly clear, the case for pardoning the 306 shot at dawn during the First World War is a moral one that makes reparations to persons still living today.

The second is an argument put forward by Ally Fogg in *The Guardian* (2013) who, like Parris, was against giving Alan Turing a posthumous pardon in 2013. Her argument invokes the problem of exceptionalism. While she illustrates her argument using the Turing case, it is also relevant to prominent cases brought by relatives of those shot for cowardice (e.g. Harry Farr). Let us first look at Fogg's argument against posthumously pardoning Turing on the grounds of exceptionalism, before extending it to the Farr case.

Alan Turing's work at Bletchley Park probably hastened the end of the Second World War by decoding German naval messages encrypted by the Enigma machine. After the war Turing was punished for being a homosexual and was forced to undergo chemical castration. The stress and shame of this could have led to his premature death in 1954, where it is suspected that he may have committed suicide (although some contest this, and believe his death was accidental). Years later many people felt that Turing was unjustly treated, and because of his heroic work as a code breaker, deserved to be singled out for a posthumous Royal pardon. Ally Fogg (2013) questions the grounds of such a pardon. In her own words: to 'single out Turing is to say all other gay persecuted men are not so deserving of justice because they were less exceptional' (Fogg 2013).

A similar kind of argument could be made in the Farr case. So, if the courts *had* decided to pardon Farr on the grounds that he was not a coward but genuinely shell-shocked, then why in particular single him out? What about the other 17 men who might also have been shell-shocked rather than cowards? Does that mean that all other shell-shocked men persecuted for cowardice were somehow less deserving and exceptional than Harry Farr? Why stop at the offence of cowardice? Were deserters also not deserving of posthumous pardoning? How about other cases?

The corollary of a Royal pardon that identifies exceptional people only is a statutory blanket pardon that draws in a variety of different cases on moral grounds. This leads full circle to the potential problem of a possible 'slippery slope' argument identified by Parris (2014), which again can be headed off by providing a strong moral case for re-evaluating the past in the present.

In sum, the most difficult problem with posthumous pardoning is not its principled conceptual defence but finding an appropriate mechanism for delivering it.

A Historical Long-View of Posthumous Punishment and Redemption

Having illuminated capital punishment, posthumous punishment and posthumous pardoning in the context of the First World War, it is helpful to look at these concepts in an altogether different historical context of understanding. Since this is a complex historical project in its own right, it is unrealistic to attempt to provide a detailed historical interpretation of capital punishment, posthumous punishment and redemption in the eighteenth century.

What is of interest here is to pick out conceptual repetition and difference over time in the ideas of capital punishment, posthumous punishment and redemption in two otherwise unrelated historical contexts.

A Bloody Code?

The capital code, or Bloody Code as it is known, saw a significant increase in the range of crimes that ended in capital punishment, with an especially wide range of property crimes being included. During the eighteenth century there was more than a quadrupling of crimes subject to the capital code (Wilf 2010). While it is true to say the capital code was bloody in the centre, especially in London, there is now strong evidence to show that it might not have been so in the periphery of the British Sate.

King and Ward, in exploring the geography and spatial dimensions of capital punishment in eighteenth-century Britain, show a widespread reluctance in areas of the periphery to implement the Bloody Code. They argue, that while it was used at the centre of the British state, it was often ignored on the periphery: in the far west, the north and the north-west of England, as well as in all of Scotland and Wales (King and Ward 2015).

Capital punishment in the British army in the context of the First World War has also been wrongly perceived as particularly blood thirsty.

Compared to the German army (which had far fewer CMs and executions) the British army's policy was more of a departure from civil law with more latitude towards summary execution. Understood from *within* the shot at dawn policy of the British army, capital punishment was used much more as a last resort for the sake of example to others. This is born out in the statistics which show that the C-in-C, Douglas Haig commuted 90% of all executions. Again, capital punishment in the form of the shot at dawn policy in the First World War was no 'Bloody Code.'

Retributive Justice, Deterrent and Posthumous Punishment

The idea of capital punishment is strongly associated with retributive justice and deterrence. Retributive justice is historically associated with posthumous punishment which has a different character depending on the particular historical context under scrutiny.

The Murder Act (1751) included the provision 'for better preventing the horrid crime of murder'. In doing so, it advocated 'that some further terror and peculiar mark of infamy be added to the punishment' (Murder Act 1751, 25 Geo. 2 c37). This involved posthumous punishment through the dissection or gibbetting of the criminal corpse. The further punishment of dissection or gibbetting was an act of symbolic power: marking the dead as infamous, while at the same time having an anonymous disciplinary effect on those that happened to witness such an event. Gibbeting, for example, involved 'the criminal body being hung raised in chains to rot and stink for any squeamish passer-by travelling along a major public thoroughfare, like the Tyburn road, to see' (Gatrell 2010). Gibbets were deliberately sited for maximum public exposure

(Dyndor 2015)—a device that expressed the will of the state and acted as a symbolic form of disciplinary power at a distance (see Foucault 1991).

The shot at dawn policy during the First World War was also a form of retributive justice that was designed as both a punishment to those that contravened military law and as an example to others who likewise might transgress regimental discipline. Being shot at dawn was also an expression of symbolic power, a deliberate way for the army to mark out deviant behaviour from the values and norms expected from a fighting soldier. The symbolic violence of execution was a form of power at a distance, one that was deliberately foisted on comrades at arms who were, either press-ganged into participating, or made deliberately aware of such killing.

Dismemberment, Disrememberment and the Execution Scene

The character of posthumous punishment changes over time and can be separated into two broad forms.

The first is a classical form of posthumous punishment which entangles disremembering the dead with posthumous dismemberment. This goes back to the time of Henry VIII in the sixteenth century, where dissection was reserved for the worst, most notorious kinds of murderers. A small number of bodies were made available through grants for the tuition of anatomy. As time passed there was pressure on the government to increase supply and, by the mid-eighteenth century, the Murder Act was passed. Here dissection was made part of the death sentence in all cases of murder, where the 'criminal' corpse supplied a growing demand of dead bodies to be used as cadavers for dissection by anatomy schools (Richardson 2001).

Posthumous punishment in the form of dissection mattered in at least three respects:

- It was feared by those with spiritual beliefs, who on sentencing would know that their dismembered body would stymie any sort of salvation of the soul in the afterlife (It was, however, certainly not feared by all). The form that this fear took depended exactly on what a belief in the afterlife entailed. For example, those with folkloric beliefs sometimes wished revenge on their tormentors, by haunting them as spirits (Linebaugh 1975; Gatrell 2010).

- Dismembering the corpse was sometimes seen as disrespectful to the memory of the dead, often angering interested spectators. It was not uncommon for family and friends to try and wrest the body from the scaffold after execution, to prevent the body being transported for dissection. (Linebaugh 1975) Dismembering the corpse often provoked anger, and ignited mob violence, especially in cases where the punishment was considered undeserved.
- Ritual dismemberment (posthumous punishment) was a form of symbolic state power—a message to others would-be criminals that certain crimes would not be tolerated. In other words 'the mark of infamy' acted as a deterrent.

The classical form of posthumous punishment officially came to an end with the Anatomy Act of 1832. The Anatomy Act ended the practice of posthumous punishment for murderers, as well as undermining the lucrative trade in the illicit supply of corpses to anatomy schools (a trade that thrived once it became clear that official supply of bodies could not keep pace with rising demand).

However, dissection of the very poor continued after the Anatomy Act since it left an avenue open for the supply of 'unclaimed' dead bodies of the very poor in society. While legislation no longer deliberately endorsed 'the mark of infamy' through dissection, it certainly didn't stamp out a much-needed supply of dead bodies for the anatomy schools. Indeed, the Anatomy Act, alongside later legislation like the Poor Law Amendment Act (1834), actively supported a supply of unclaimed paupers body's from workhouses—a fate that caused much fear and angst amongst the poor and destitute in society (Hurren 2012). It could be argued that the nineteenth century state was still actively involved in posthumously harming the very poor, through an act of collusion with institutions that supplied much needed corpses and their parts, rather than as an overt state agent that prevailed over retributive justice and posthumous punishment of criminals.

The idea of posthumous punishment post capital punishment is reinvented in the twentieth century through capital punishment and deliberately dishonouring a good soldiery reputation. The big difference between posthumous punishment in its classical form following the Murder Act (1751), and posthumous punishment in the twentieth century—by way of being dishonourably shot at dawn in the First World

War for example—is that there is no intentional desire to disremember the dead long-term, and certainly no dismemberment of the body and denial of funerary custom. Those 'shot at dawn' were not intentionally vilified by the army beyond being set up as 'bad' fighting men for the sake of example and regimental discipline. Otherwise their relatives were treated with some compassion by the army and their corpse left intact to be buried in the normal way. Long after the war had ended, the different national commemoration cultures honoring the fallen in the war rarely mark out those executed by their own comrades from those fallen in battle.

The weak comparison between eighteenth-century hangings and executions witnessed in the First World War is that they both provoked mixed reactions in those who observed them, largely depending on how witnesses felt about justice served. This said, the character of the execution scenes in these two historical contexts could not have been more different in character. Hangings at Tyburn and Newgate during the eighteenth century attracted large voluntary voyeuristic and ambivalent crowds with executions sometimes resembling a carnival-like atmosphere (Gatrell 2010). By contrast executions during the First World War were sombre disciplined affairs, comrades in arms being forced to witness and/ or participate in a dawn execution (Johnson 2015).

Redemption and Posthumous Pardoning

If the idea of capital and posthumous punishment has long historical rhizome-like roots that repeat with a difference over time then so does the notion of posthumous pardoning.

In the UK, the power to grant pardons and reprieves is known as the Royal prerogative of mercy—a Royal prerogative is where a King or Queen can grant a pardon to a convicted person. This is also called a Royal pardon. Historically, the Royal prerogative of mercy (or Royal pardon) was an absolute power wielded by the monarch alone to pardon an individual for a crime, but the power for the monarch to use this was significantly curtailed by the end of the seventeenth century (Loveland 2009). In more recent, times this power has been delegated to the judiciary and sovereign ministers (Leyland and Anthony 2007).

Reprieve and pardon from capital punishment after the Murder Act (1751) was possible at any time, at the last moment, just before hanging, and even after hanging (if a criminal miraculously survived the hangman's noose). In the case of the latter, criminals that survived might have had their sentence commuted to transportation to the colonies. This was an early form of pardoning; where divine mercy was respected by a 'posthumous redemption' of sorts. Criminals that survived the noose were offered commuted sentences so long as the attempted execution was legal. In such cases they had undergone a socially symbolic or legal death. Of course, legally killing a condemned man does not always end in their physical death, which on occasion led to the curious situation of a criminal being 'legally dead' whilst remaining physically alive (Hurren 2016).

The purpose and function of a pardon has changed over time. Alongside the few Royal pardons since the Second World War that have been mobilised to save and free people, there is the rise of the posthumous pardon. Whereas before this time a pardon was most commonly executed as an act of clemency, to physically save individuals from the full force of the law, it has, in contemporary times, increasingly been used as a way of rehabilitating individual identity and reputation after death.

Summary

This empirical chapter has examined capital punishment, posthumous punishment and pardon.

This has been illustrated through a multiplicity of perspectives, largely through notions of capital punishment during the First World War and the debate around posthumously pardoning soldiers executed by being 'shot at dawn'. In order to move the debate on, the author has concentrated on reframing posthumous punishment and pardoning on more conceptual grounds, by considering what punishment and pardoning was for.

Posthumous punishment and pardoning has also been briefly understood from a long historical view, such ideas repeating with a difference over time.

References

Barton, P. (2016). *From both sides of the wire, end game—Episode 3*. Retrieved August, 2016, from http://www.bbc.co.uk/programmes/b07n22gn.

Carr, E. H. (1990). *What is history?*. London: Penguin.

Corns, C., & Hughes-Wilson, J. (2002). *Blindfold and alone*. London: Cassell.

Dyndor, Z. (2015). The gibbet in the landscape: Locating the criminal corpse in mid-eighteenth century England. In R. Ward (Ed.), *A global history of execution and the criminal corpse*. Basingstoke: Palgrave Macmillan.

Fenton, B. (2006). Pardoned the 306 shot at dawn for 'Cowardice'. *The Telegraph*. Retrieved October 4, 2016, from http://www.telegraph.co.uk/news/1526437/Pardoned-the-306-soldiers-shot-at-dawn-for-cowardice.html.

Fogg, A. (2013). Alan turing pardon is wrong. *The Guardian*. Retrieved October 4, 2016, from http://www.theguardian.com/commentisfree/2013/dec/24/alan-turing-pardon-wrong-gay-men.

Foucault, M. (1991). *Discipline and Punish*. The Birth of the Prison St. Ives: Penguin

Gatrell, V. A. C. (2010). *The hanging tree: Execution and the English people 1770–1868*. Oxford: Oxford University Press.

Gosling, G. C. (2014). *Musings, 'Shot at Dawn'*. Retrieved October 4, 2016, from https://gcgosling.wordpress.com/2014/01/13/shotatdawn/.

Hansard (2006, September). col 135S. Retrieved June 7, 2017, https://www.publications.parliament/pa/cm200506/cmhansrd/vo060918/wmstext/60918m0001.htm.

Haraway, D. (2004). Situated knowledges. In S. Harding (ed.). *The Feminist Standpoint Theory Reader*. London: Routledge

Holmes, R. (2004). *Tommy: The British soldier on the western front, 1914–18*. London: Harper Collins.

Hurren, E. T. (2012). *Dying for Victorian medicine: English anatomy and its trade in the dead poor, c. 1834–1929*. Basingstoke: Palgrave Macmillan.

Hurren, E. T. (2016). *Dissecting the criminal corpse: Staging post-execution punishment in early modern England*. Basingstoke: Palgrave Macmillan.

Johnson, D. (2015). *Executed at dawn: British firing squads on the western front 1914–18*. Stroud: The History Press.

Kentlive. (2010). *Battle continues for descendants of deserter*. Retrieved October 5, 2016, from http://www.kentlive.news/battle-continues-descendants-deserter/story-11997556-detail/story.html.

King, P., & Ward, R. (2015). Rethinking the bloody code in eighteenth century Britain: Capital punishment at the centre and on the periphery. *Past and Present, 228,* 159–205.

Leyland, P., & Anthony, G. (2007). *Textbook on administrative law* (6th ed.). Oxford: Oxford University Press.

Linebaugh, P. (1975). The tydburn riots against the surgeons. In D. Hay et al. (Eds.). *Albion's fatal tree*. London: Pantheon.

Loveland, I. (2009). *Constitutional law, administrative law and human rights: A critical introduction* (5th ed.). Oxford: Oxford University Press.

The Murder Act. (1751). (25 Geo.2 c.37).

Norton-Taylor, R. (2006, August 16). Executed WW1 soldiers to be given pardons. *The Guardian*. Retrieved October 5, 2016, from http://www.theguardian.com/uk/2006/aug/16/military.immigrationpolicy.

Parris, M. (2014). Why I'm against posthumous pardons, even for alan turing. *The Spectator 26 July 2014.* Retrieved October 5, 2016, from http://www.spectator.co.uk/2014/07/why-im-against-posthumous-pardons-even-for-alan-turing/.

Putowski, J., & Sykes, J. (2014). *Shot at dawn: Executions in world war one by authority of the British Army act.* Barnsley: Pen & Sword.

Richardson, R. (2001). *Death, dissection and the destitute.* London: Phoenix Press.

Taylor-Whiffen, P. (2011) Shot at dawn: Cowards, traitors or victims? *BBC History.* Retrieved October 7, 2016, from http://www.bbc.co.uk/history/british/britain_wwone/shot_at_dawn_01.shtml.

Wilf, S. (2010). *Law's imagined republic: Popular politics and criminal justice in revolutionary America.* New York: Cambridge University Press.

Open Access This chapter is licensed under the terms of the Creative Commons Attribution 4.0 International License (http://creativecommons.org/licenses/by/4.0/), which permits use, sharing, adaptation, distribution and reproduction in any medium or format, as long as you give appropriate credit to the original author(s) and the source, provide a link to the Creative Commons license and indicate if changes were made.

The images or other third party material in this chapter are included in the chapter's Creative Commons license, unless indicated otherwise in a credit line to the material. If material is not included in the chapter's Creative Commons license and your intended use is not permitted by statutory regulation or exceeds the permitted use, you will need to obtain permission directly from the copyright holder.

CHAPTER 5

Posthumous Harm in the History of Medicine

Abstract Posthumous harm, in the first instance, is understood in the context of Alder Hey Hospital in Liverpool, where in the late twentieth century improper procurement and retention of organs and other human materials, supposedly under the auspices of medical research, occurred against the express wishes of the next of kin.

The notion of posthumous harm is first explored more formally, through a public inquiry, which found that institutional failures aggravated the original harm of the improper removal and retention of organs. It is then explored more deeply, through the parental oral evidence to The Royal Liverpool Children's Inquiry Report (2001b). From the perspective of grieving family and friends, posthumous harm is constitutive of: a breach of proper consent; an assault on grief and memory; as well as a contravention of religious belief and funerary custom.

The counterpoint to posthumous harm is the notion of posthumous redemption. From the point of view of the medical professional, the cadaver is a redemptive force in medicine; it is a resource for saving lives (cadaveric organ donation), medical research and medical education and training. While the intention behind the improper post-mortems at Alder Hey was medically redemptive, organs and tissues were stored away and largely unused.

Posthumous harm and redemption are finally understood from the historical long-view. This chapter ends with a comparative view between body-snatching in the Georgian period and 'organ-snatching' at Alder Hey two hundred odd years later.

Keywords Alder Hey · Body-snatching · Organ-snatching · Posthumous Harm · Redfern · Symbolic Harm and Violence

Contemporary Perspectives on Posthumous Harm and Redemption: Alder Hey

An Overview of Events

One of the most infamous examples of posthumous harm in contemporary times arises out of the organ retention scandal at Alder Hey Hospital in Liverpool in the 1990s.

Awareness of the retention of organs at Alder Hey arose from an altogether separate public inquiry into the unusually high infant mortality rate after cardiac surgery at the Bristol Royal Infirmary. On 7 September 1999, the heart specialist Professor Anderson of Great Ormond Street gave evidence to the Bristol Inquiry. In his evidence, Anderson pointed out the advantages of post-mortem retention of hearts for research and teaching purposes. In particular he mentioned the impressive collection of hearts held at Alder Hey Children's Hospital, which dated from 1948 (Hall 2001; Harrison et al. 2003, p. 49).

Anderson's observation was picked up by the local media on 18 September 1999, when two days later Ms. Hilary Roland, the Chief Executive of Alder Hey, gave assurances that the retention of organs was not dissimilar to that of other hospitals. The news caused enquiries from potentially affected parents of deceased children wanting to ascertain whether their children's organs had been retained. (Harrison et al. 2003 p. 49) After further investigations by the hospital management it was found that Ms. Rowland's initial assurances had been premature when it was discovered that the Professor of Pathology, Dick van Velzen, had authorised the retention of multiple organs, tissue fragments and whole foetuses, mainly in the basement at Myrtle Street.

Many parents were affected: 2080 organs had been removed and retained from 800 children and stored in pots. In addition, 1500 foetuses were also discovered in storage—either miscarried, stillborn, or aborted without consent (Batty 2001). For a single hospital, the retention figures were huge.

Parents wanted to know whether their children's organs had been retained, and wanted them located and returned for reinterment. The hospital failed to meet parental demands. The general feeling amongst parents was that the hospital was mismanaging information and or deceiving them as to what had happened to their children. In response, a support group was formed: Parents who Inter Their Young Twice (PITY2). This provided a self-supporting environment and a more organised and effective voice in dealing with hospital management.

On December 3, 1999 the new coroner for Liverpool suggested that the retention of organs had been unlawful, heightening parental concern and anxiety and launching Alder Hey back into the media spotlight once more. This and the *prima facie* evidence of wrong-doing and mismanagement culminated in the government announcing an independent Public Inquiry chaired by Michael Redfern QC in early 2000. Meanwhile the relationship between the hospital and the affected parents continued to worsen: parents complaining that the hospital was misleading them into believing they had buried their children intact when in fact they had not. Instead, most of the human material from the children of affected parents was stored unused (Harrison et al. 2003, p. 50).

The Alder Hey organ retention scandal was thoroughly investigated through a public inquiry. The results of which were reported on at length in The Royal Liverpool Children's Inquiry Report by Michael Redfern QC. This was published in January 2001. The Redfern Inquiry (2001), or simply 'Redfern' as it will be referred to from here on in, reported on the worst organ removal and retention scandal in Britain. When talking about Redfern, it is worth distinguishing between the summary and recommendations of the Inquiry Report (2001a) and the full Inquiry Report (2001b).

Redfern provides a helpful window into understanding:

- Problems with the practice of improper procurement, retention and storage of human material. This is discussed in a summary of Redfern's formal conclusions. It also provides recommendations of how it might be remedied;
- Why the posthumous harm of improper procurement, retention, storage and disposal affected parents so deeply. Much of the material about parents' emotional reaction in the parental oral evidence is available from the full Inquiry Report. This evidence provides insight into parents' inner lives and why posthumous harm matters to the next of kin in the first place. This is discussed in the parental oral evidence to Redfern.

A Short Summary of Redfern's Formal Conclusions

The Redfern Inquiry summary and recommendations provide an incisive formal account of what went wrong at Alder Hey. While the summary begins with the misconduct of the head of pathology, the failures are institutionally endemic.

The Misconduct of Persons: Professor Dick van Velzen

Much of the furore around the retention scandal focused on the activities of Professor Dick van Velzen, the Head of the Foetal and Infant Pathology Unit at the University of Liverpool, and honorary paediatric pathologist at Alder Hey from 1988 to 1995.

It was clear from the very start that van Velzen's activities as a pathologist were divergent from the norm. Within a week of taking up his position, van Velzen issued the instruction that there was to be no disposal of human material. He wanted 'every organ removed in every case' (Redfern 2001a, p. 8). Before van Velzen, pathologists had only retained sections of organs and the heart, lungs or brain in relevant cases. Naturally the store of human material started to grow to support his research interests, which he justified on the basis of developing a resource which he could exploit depending on the direction his research developed in the future.

Van Velzen's professional misconduct was extensive and is thoroughly documented in Redfern (2001a, 2001b).

What follows is an illustrative and non-exhaustive list of what Redfern found van Velzen guilty of:

- lying to patients generally and lying to them about his post-mortem findings more specifically;
- deceiving both Alder Hey and the university;
- unethical and illegal retention of organs;
- falsifying research applications, post-mortem reports and encouraging staff to falsify records and statistics;
- ignoring consent that stated a preference for limited post-mortems;
- failing to keep proper records of stored organs, and failing to maintain proper accounting procedures (Redfern 2001a, pp. 9–10).

Redfern's recommendations were designed to stop malpractice by rogue individuals like van Velzen. This was largely achieved by augmenting a trust in systems; where trust at the level of the institution is about accountability and 'super' accountability that can no longer afford to take the risk of employing untrustworthy persons (Pilgrim et al. 2011).

Redfern marked a shift from a paternalistic culture which stressed a blind trust in individuals to trust in institutions and institutional procedures and systems where hospitals had to be more accountable for: whom they employed; how they openly and honestly dealt with serious incidents as they arose; and how they implemented consent and reporting procedures (Harrison et al. 2003; Pilgrim et al. 2011).

Relationship Between the University and the Hospital

Michael Redfern remarked on the uncooperative relationship between the University and Alder Hey, culminating in the University distancing itself from the hospital once news of the retention scandal broke. He observed how this difficult relationship between the two institutions provided an opportunity for van Velzen to play one institution off against the other in support of his own agenda. Redfern concluded that van Velzen's worst excesses might have been prevented had the institutions had a better relationship. This prompted Redfern to recommend that institutions with dual clinical and academic functions had to develop relationships that 'fostered good faith in both directions' (Redfern 2001a, p. 15).

The Role of the Coroner

Michael Redfern discovered multiple failings. Clinicians were not always sure under what circumstances death had to be reported to the coroner. The coroner sometimes wrongly delegated post-mortems to the coroner's office. The coroner also did not follow up requests on histology with van Velzen, exacerbating the incompleteness of post-mortems and the illegal retention for research purposes. Some clinicians had been abusing the system by threatening parents with mandatory coroner's post-mortems in order to put pressure on them to agree to a voluntary hospital post-mortem (Redfern 2001a, p. 4). He concluded that failures in understanding the coroner's role contributed to a delay in identifying van Velzen's malpractice.

The role of the coroner needed to be clearly established. Redfern's recommendations spelled out the role of the coroner and a need for clinicians to be educated in proper procedures that involved the coroner's office (Redfern 2001a, pp. 18–20).

Serious Incident Procedure and Record Keeping

After thoroughly looking into van Velzen abuses and the conditions which exacerbated them, Redfern established that the hospital had inadequately disseminated the news of improper retention to parents.

Alder Hey failed to provide honest face-to-face communication of the news of organ retention. It also failed to provide bereavement counselling and support for affected families (Redfern 2001a, p. 12).

The management of the retention news was far from adequate. News was drip fed. Information was often inaccurate. News of retention was commonly delivered insensitively (Redfern 2001a, p. 12). All of which greatly aggravated the original harm caused by improper retention. Many families affected had parts of their dead children returned to them on a piecemeal basis for reinterment over unacceptable periods of time, which further contributed to their suffering post-retention news.

Redfern recommended that Trusts introduce serious incident procedures (Redfern 2001a, p. 13). This involved appropriate forms of communication with those affected, who were to be communicated with in an open, honest and sensitive way. In order to handle the special sensitivity of such news it was recommended that bereavement experts be involved as a matter of course. Redfern also recommended overhauling the pathology record system, so that receipt use, and ultimate disposal of organ and tissue sample could be tracked at all times (Redfern 2001a, pp. 13–14).

The Issue of Consent

The Human Tissue Act 1961 (HTA) required clinicians to establish whether, after reasonable enquiry, they had any reason to believe that surviving relatives 'objected' to their kin being used for therapeutic purposes, medical education or research.

There is overwhelming evidence to show that the requisite demands of the HTA had not been met. In other words, clinicians had not made 'reasonable enquiries' to ascertain whether parents 'objected' to postmortem procedures, sometimes even ignoring their wishes and putting pressure on them to change their minds. Whilst Michael Redfern recognised the role that paternalism played in this, the Inquiry Report keenly points out that paternalism did not explain it away. Redfern points to the fact that clinicians 'lacked any proper consideration of the Act in the first place' (Redfern 2001a, p. 3).

Redfern's aim was to make the consent more procedurally rigorous and a less conceptually ambivalent process. There was also a need to train and educate. Doctors needed educating and training in the taking of consent, and the public needed to understand why organs needed to be retained for medical research and educational purposes.

At the heart of Redfern's recommendations was a conceptual shift in the idea of consent (Redfern 2001a, pp. 23–24). Having a more robust law around consent was a way of changing the focus from clinical- to patient-centred interests.

The HTA was still a form of presumed consent because 'reasonable enquiry' to confirm 'no objection' carries with it a significant element of clinical presumption that post-mortems are acceptable. Redfern favoured a shift towards 'fully informed consent'. This moved the responsibility of decision-making away from the doctor and on to the parent or next of kin. This significantly improved consent in two ways.

Firstly, informed consent challenges the paternalistic attitude that 'doctor knows best'. This presumption was especially dangerous when it was assumed that the next of kin would be too distressed to discuss consent regarding a post-mortem.

Redfern identified paternalism as a deeply contributing cultural factor to the retention tragedy. Redfern took the view that not involving the parents in taking responsibility for post-mortem retention would only serve to increase future distress. Involving them, as the parental evidence suggests, might relieve the anger, resentment and guilt at having the decision taken away from them—the one caveat being that involving the parents in decision-making had to be done sensitively, allowing time and space for them to reach a stable decision that they could live with.

Secondly, informed consent also militates against clinicians from acting in their own *self*-interest. The reality at Alder Hey was that clinicians were often motivated by their own clinical or research interests. They were regularly blind to a deep parental need to have the corpse treated with respect, so that the integrity of the dead might be preserved wherever possible in line with the demands of normal grieving, religious belief and funerary custom.

Beyond the Formal Conclusions of Redfern
It is mistaken to think of the posthumous harm of retention too narrowly, as if it might somehow be reduced to a series of formal conclusions in a public inquiry. There are four mistakes that can be made.

The first mistake is to think that improper removal, retention and storage are the only issues that matter concerning posthumous harm. Alder Hey represents a breakdown of trust in healthcare and can be reframed in such terms (Harrison et al. 2003; Pilgrim et al. 2011).

The second mistake is to think of retention at Alder Hey as an isolated occurrence. Liam Donaldson, the Chief Medical Officer, reported in February 2001 that improper retention and disposal of organs was widespread in the UK. Alder Hey got most of the bad press, and was probably only unique in the scale of retention and in the furore it caused. Importantly, Donaldson discovered that other NHS institutions around the UK were also guilty of flouting the HTA and improperly removing and retaining organs:

> ...elsewhere in the NHS it is clear that organ retention without relatives' full knowledge and agreement was widespread. The recent national summit on organ retention organised by the Chief Medical Officer, Professor Liam Donaldson, confirmed that this was also the experience of parents in many other parts of the country. Professor Donaldson's census shows that 105,000 organs are retained across the country. Poor standards of cataloguing and record-keeping mean that these figures may not be wholly accurate. Twenty-five thousand hospitals account for 88% of the organs. At least 16,500 of these organs and tissues have been retained in apparent contravention of the law because they came about as a result of coroners' post-mortems where the organs should not have been kept beyond the time needed to establish cause. (HL Deb. 2001, 621 c. 574)

The third mistake is to become overly focussed on the formal conclusions—especially the summary and recommendations of the Inquiry.

The summary and the full report generally concerns what went wrong at Alder Hey from an institutional point of view. An over-focus on institutional failure can lead to an inability to understand the impact of posthumous harm on its victims. For this we need to understand the more implicit and informal parental evidence to Redfern available from the full Inquiry Report.

The fourth mistake is to think of Alder Hey as historically unique. It is not. Lessons can be learned from the past, as well as from a public inquiry.

The idea of posthumous harm, it will be argued, repeats with a difference over time. Before looking into this in some depth towards the end of this chapter, and making historical comparisons with unconsented dissection in the nineteenth century, we need to understand the impact of posthumous harm through parental evidence to Redfern.

Understanding the Parental Oral Evidence to Redfern

The parental evidence falls into two broad analytical categories:

- A surface level of analysis in which what parents say to the Inquiry is fairly self-evident in respect to why it is harmful. This is illustrated by issues around consent and the spectrum of deceit;
- A deeper level of analysis where what parents say to the Inquiry is not so self-evident, but nevertheless gives important clues as to why posthumous harm matters to them. This is illustrated by issues around identity beyond biological death and how harm to memory and biography through dismemberment is possible in a narrative sense.

Consent and the Spectrum of Deceit
Parents were deceived by improper removal and retention of organs, either through the initial consent procedure or, by the hospital failing to inform and support parents properly afterwards. There is a whole spectrum of deceit evident through the parental evidence presented to Redfern.

At one end of the spectrum parents were lied to. For example:

Kathryn – 15 years Kathryn developed Hodgkin's disease and died at Alder Hey Hospital in 1993 … On 8 December 1999, her parents were informed by Alder Hey that Kathryn's heart, chest and abdomen had been retained. On the 20 December 1999 they had received a letter from Ms Hilary Rowland, Chief Executive at Alder Hey, indicating the heart, lung, liver, liver, spleen and kidneys had been retained. In the post-mortem report Prof van Velzen said that only a small mid sternal incision approach was made with splitting of the caudal sternum. Only the upper organs and lower aspects of the chest organs were brought into view and inspected… Only organ biopsies were taken. This was a fiction confirmed by the list of organs described by Ms Rowland in her letter of 20 December 1999. (Redfern 2001b, pp. 396–397)

Being deliberately lied to by clinicians like van Velzen was compounded by Alder Hey's inept handling of events. Parents had information withheld from them and were sometimes given contradictory information about retention. For example:

> *Anthony – 3 years 10 months* Anthony was born with congenital heart disease. He died in 1996 within 24 h of major heart surgery... In late September 1999 they [the parents] contacted Alder Hey and although they were being treated professionally they always felt that certain information was being withheld. The hospital was evasive in certain areas... At first they were told that the heart had not been taken. As an afterthought they asked what tissue samples had been taken and to their surprise were told brain, stomach, one kidney and one lung. They were then contacted by the treating clinician who told them the heart had been taken as well. (Redfern 2001b, p.422)

Some parents interpreted the improper retention of their children's organs without consent as theft. For example:

> *Tony – 11 days* Tony died in 1994. His precise cause of death is still under investigation. They [the parents] were told that there would be Coroner's post-mortem examination... When they rang Alder Hey in late 1999 to enquire if any organs had been retained they were told the following day that there had been retention. They asked what had been taken and the reply was 'everything basically'... His mother told the hospital that they had *stolen* the organs and she wanted a 100 percent guarantee that Tony's were not being retained, to which the hospital said 'alright you have got a 100 per cent guarantee' after previously refusing to give such a guarantee... (Redfern 2001b, p. 412 [author's italics])

Clinicians often did not explain the consent procedure and/or parents were too distressed to be able to give consent due consideration. Clinicians also readily conflated the meaning of taking tissue with organs, misleading parents into exactly how their children had been handled post-mortem and what had and what had not been taken. For example:

> *Ross – 5 months* was born prematurely at 27 weeks and died in 1990 at Liverpool Maternity Hospital... His parents consented to a post-mortem examination to determine the cause of death if it would help other children who had the same disease. No steps were taken to explain the consent

form to them. Because of their distress they describe signing it 'blind'. They realised that small samples would be taken from organs in the post-mortem procedure but understood that to mean a small piece of tissue for microscopic examination. They were never told whole organs would be removed and retained... They thought they buried their son intact whereas in fact they buried a husk. (Redfern 2001b, p. 399)

A mother felt that Alder Hey had deliberately used 'tissue' as a euphemism in order to remove whole organs. For example:

Ryan – 19 days Ryan was born with congenital heart disease. He died following open-heart surgery in 1995 at Alder Hey... They feel they used the word 'tissue' when it suited them, and if they were looking to have an organ from a child and put it another child that would have been organ donation but because the organs have been retained for medical research purposes they are then classed as tissue samples. The mother is unhappy at this false distinction. (Redfern 2001b, p. 432)

Personal Identity and Its Continuation Beyond Death
One commonplace misconception is that an individual's identity ends at biological death and is no longer a relevant 'fact' post-mortem. For example:

Sam – 18 months Sam was born with congenital heart disease. He died in surgery in 1990 at Alder Hey. A coroner's post-mortem examination was carried out. The post-mortem examination was not explained to his parents... They received news of organ retention in January 2000. Eventually they were told there had been full retention... *The impression given by Alder Hey was that an individual's identity ends at a post-mortem examination if not death...* (Redfern 2001b, p. 425 [author's italics])

Death from a clinician's perspective tends to revolve around the notion of biological death only. Their *raison d'être* is to save life and do no harm. The desire to save life at Alder Hey went hand in hand with a caring attitude to parents of sick children.

Several Alder Hey parents observed the contrast between the high quality of care their children received in life compared to in death (Redfern 2001b). The lack of care towards the dead by clinicians at Alder Hey was compounded by their training. In the words of a nurse communicating the news of organ retention to a parent: 'try not to look at this emotionally it is just tissue' (Redfern 2001b, p. 414). While factually true at the level of biological death, it shows a lack of empathy and understanding of grief when care of the recently deceased naturally continues.

Posthumous Harm as Narrative or Symbolic Harm to the Dead
At a more conceptual level of understanding, the corpse is both a physical unity and a locus of symbolic meanings. For parents the corpse of their recently deceased child is imbued with intimate memories and associations. For a pathologist, on the other hand, the corpse as a cadaver accrues objective and scientific understanding of the cause of death. Ideally, of course, the clinician needs to understand both perspectives: her own scientifically motivated one, as well as that of the next of kin, who are still emotionally attached to the deceased. This flexibility of perspective was lacking at Alder Hey.

One of the most striking pieces of evidence to Redfern describes how a mother's memory of her deceased child has been 'ruined' by inappropriate removal and retention of his organs. For example:

> *Kenneth – 5.5 weeks* Kenneth died in 1987… In December 1999 his mother contacted Alder Hey and two weeks later was told that the heart only had been retained… The family has been bitter at the discovery of heart retention… *She says that the memory of her child has been ruined by living under the illusion that he was buried intact when in fact he was missing his heart. She cannot even look at pictures of him now because she just sees him in a different way.* (Redfern 2001b, pp. 410–411 [author's italics])

From the perspective of Kenneth's mother, the removal of Kenneth's body parts without her permission was perceived as posthumous harm. That is, body parts like the heart, eyes, hands represent relationships with significant others symbolically inscribed. So, if body parts are missing, such as the heart, then harm has been done, because body parts represent the memory of a deeply personal relationship (Dickenson and Widdershoven 2001, Tomasini 2009). The improper retention of Kenneth's heart affected his mother's capacity to remember him as he was. This is probably exacerbated in the case of babies and infants, where the memory of that child is yet to be properly established. So any inappropriate dismemberment is going to affect the particular association and fragile preciousness of any early biographical memory acquired. It is also harm to hope and future life; that is, the hope that parent holds for that child's future and the memory of an imagined and anticipated life together with that child as they grow up in a family.

The idea of posthumous harm for those parents affected by Alder Hey 'complicated grief'. Complicated grief constitutes both symbolic and intinsic harm. The symbolic harm through desecration of the body is a vector for a more complicated and psychologically distressing grief.

Complicated grief involves parental guilt at not protecting their child from harm post-mortem. That is to say, the clinical deception that allowed clinicians to remove organs without proper consent made parents feel guilty that they had not 'protected' their loved ones in death. For example:

> Sam 18 months: "They feel that they protected their child in life, but *in death when he needed their protection more than ever, they feel guilty they let him down in allowing or permitting organ retention.*" (Redfern 2001b, p. 425 [author's italics])

From a biological view of death this is quite unintelligible as no harm can befall the dead, so why protect them?

Perhaps this is why some clinicians at Alder Hey seemingly lost interest in the dead, not showing the same amount of care to recently deceased children, when every effort was extended to save their life. Unsurprisingly, many clinicians are so conditioned into saving lives that *some* may overlook the needs of grieving parents and their overwhelming desire to 'protect' their dead.

From the perspective of social death 'protecting' the dead is perfectly intelligible as a recently deceased person retains a narrative identity; that is, body parts are symbolically inscribed with interpersonal memories, so if those parts are damaged or go missing, parents feel guilty at not 'protecting' the memory of their child.

This idea of the 'continuing bond' between living and the dead is well established in the literature (Klass et al. 1996). The 'continuing bond' needs to be protected in both a literal and figurative sense. Failing to protect the physical integrity of the dead potentially affects memory and their narrative identity after life. In short, inappropriate dismemberment amounts to symbolic harm.

The symbolic harm of dismemberment was further complicated for parents who had strong religious beliefs. The misplaced paternalism and insensitive attitude at Alder Hey often deeply affected parents, denying them their need to take control of the funeral and bury their children whole according to, for example, their Catholic religious beliefs and funerary customs. For example:

> *Christopher – Stillborn* Christopher (stillborn) Christopher was born prematurely stillborn in 1987. He was taken away at birth... The hospital insisted that the baby be buried in hospital grounds. They [the parents] were told that it would be a dignified ceremony. They asked if they could see their son buried. *They were told that it would not be possible and that seeing the burial grounds would only upset them. They were told to have another child.* The parents had no control over what happened to Christopher. They wanted to bury him themselves. *They are a Catholic family and burial is important to them. To bury their child intact is part of their religious belief...* (Redfern 2001b, p. 430 [author's italics]).

Not only were Christopher's parents dealt with extremely insensitively, to the point where clinical patriarchy borders on being callous, but the behaviour showed no understanding of the funeral ritual and the part this plays in the religious beliefs that the parents had.

From the formal perspective of Catholic doctrine dismemberment *could* be interpreted as future-orientated symbolic harm. That is, Catholic doctrine encourages belief in the integrity of human remains for burial in consecrated ground for the sake of resurrection at the Last Judgment. The implication is that to deny resurrection of the body is to deny the Resurrection of Christ (Burke 2016).

This, of course, does not explain why exactly Christopher's Catholic parents wanted to bury their child intact. While the evidence needs to be taken at face value, such Catholic doctrine at least provides a clue as to why not interring a person whole might provoke anxiety and fear amongst devout Catholics.

Posthumous Redemption Narratives: Failures and Successes
Because of the harm perpetrated by the organ retention scandal at Alder Hey, it is easy to overlook the redemptive narratives that may have motivated clinicians in their wrongful attempts to remove and retain organs post-mortem.

Anonymous body parts have a redemptive narrative in the history of medicine, either directly, through saving another's life through cadaveric organ donation, or indirectly by helping science understand the cause and course of disease. Retention of organs and tissue from cadavers would also have provided a resource in the teaching of medical students.

Looking more specifically at the Alder Hey case, the taking of organs post-mortem could have had a redemptive affect if parents had been asked properly and the organs had been put to some use in saving other lives in some way.

Richardson has claimed that parents might well have given their permission if they had been asked properly (Richardson 2001, p. 416). From the oral evidence to Redfern, some parents were clearly overcome by the shock of loss, and without any bereavement support or any time to process the information many made decisions that they came to regret and feel guilty about later (Redfern 2001b, pp. 388–434). Others, who did consent to limited post-mortems, expressed preferences in their desire to help others in some practical way. The narrative of a child's life continues on after life, in 'the gift of life'—where a donor's organs may save the life of sick child awaiting a transplant. The importance of 'the gift of life' is well established in the literature on attitudes to donation. Indeed, the 'sacrifice' of a parent giving up *their* dead for donation or research purposes is outweighed by the prospect that it may be understood as a 'gift of life'—helping another sick child in some way (Sque et al. 2006, pp. 117–131). Put another way, no parent affected by the Alder Hey retention scandal gave permission for their children's organs and tissue to stored and left unused in the basement at Myrtle Street.

If the redemptive act of saving another life through organ retention had some potential—both as an intrinsic fact in the preserving biological life where it was failing, and as a symbolic one in continuing the narrative of the life of an ailing donor—much of this was a wasted opportunity at Alder Hey, where organs and tissue piled up, unused, in storage.

Interestingly van Velzen was interviewed in 2001 (Dickson 2001) about his role in the whole debacle. He maintained that his motivation in removing organs wholesale 'was demanded by standard international protocols' and that the growing collection was to be used for future research purposes yet to be determined.

He claimed that the 'organs piled up' when the money ran out and that he no longer had the support he needed to keep on top of his day-to-day clinical case load. At no point did he say he was wrong or did he say he was sorry (Dickson 2001).

It is difficult to give much credibility to van Velzen's justifications. What is more probable is that van Velzen was in what Sartre would call 'bad faith' (Sartre 2005) about his role in the retention scandal. While van Velzen was not the only person that deserved to shoulder blame, he seems to have been afflicted by a 'moral blindness' years after the facts had come to light.

The only practically redemptive narratives which stand out at Alder Hey are lessons learned from the tragedy itself. Indeed, the Bishop of Liverpool, who paid tribute to the victims of Alder Hey in a special service for its victims proved prophetic. In his address he thanked 'God for the parents' courage and restraint.' He 'promised that their children's deaths and their own sufferings had not been in vain: their courage to confront and expose illicit behaviours meant that things would be different, and better, in the future' (cited in Richardson 2004, p. 45).

This was prophetic. Redfern paved the way for many radical and helpful changes to safeguarding against posthumous harm of this kind. It also provided a cornerstone for a radically revised HTA (2004) that introduced the notion of fully informed consent procedures. This not only affected how post-mortems were conducted, it signalled how other procedures were conducted in the NHS, paving the way for a wider cultural change away from medical paternalism and towards a more patient-centred approach where responsibility is more equally distributed.

A Historical Long-View of Posthumous Harm and Redemption: Alder Hey

The character of posthumous harm repeats with a difference over time. It is not a notion that is historically unique to Alder Hey. To understand how the character of posthumous harm over time repeats with a difference, it is helpful to offer a comparative account: body-snatching in the Georgian period versus organ-snatching in the late twentieth century at Alder Hey.

A Historical Long View of Posthumous Harm: Comparing Body-Snatching to Organ-Snatching

Improper Procurement and Retention

Taking organs of dead children without parental permission at Alder Hey is a practice *The Economist* (2001) dubbed the 'return of the body-snatchers'. There is a historical parallel to be drawn between the practice of body-snatching in the Georgian period and 'organ snatching' at Alder Hey some two hundred or so years later.

As regards the law both body *and* organ 'snatching' were illicit rather than straightforwardly illegal practices. The removal of corpses from graves by the 'resurrection men' was not illegal before the Anatomy Act, although stealing from the corpse and or 'knowingly' dissecting the corpse was. It would be more precise to say that body-snatching was an 'extra-legal' activity (Richardson 2006, p. 155).

The removal of organs at Alder Hey for the most part was also not (straightforwardly at least) illegal. The pathology team at Alder Hey seemingly went through the legal motions of securing consent for post-mortems from parents of dead children. After the scandal broke, the retention issue was deemed unlawful and parents secured a successful legal challenge against the removal of their children's organs.

Even though 'organ-snatching' did not involve anything as dramatic as stealing a corpse from a grave in the dead of night, it did, under the cloak of seeming medical respectability, amount to something similar to body-snatching. That is, from the perspective of family and friends the removal of organs without parental permission was sometimes experienced as an act of theft. In the case of body-snatching, bodies of the recently deceased were stolen out of graves to the chagrin of those watching over them.

The Commodity Value of the Cadaver

The motivation behind organ-snatching and body-snatching is similar: that is, human material post-mortem has a strong commodity value, even though what is valuable about the human corpse changes over time.

The transformation of the cadaver to an object of trade—commodification—took off in the body-snatching era, when the gallows provided nowhere near enough bodies required for teaching and research purposes. Demand outstripped supply and opened up a lucrative market for human remains to anatomy schools (Richardson 2001a, pp. 52–72).

The corpse represented monetary value to the body-snatcher. From the perspective of the anatomist, the corpse is a cadaver, meaning a dead body intended for dissection. The change of signification of corpse to cadaver for dissection and anatomization is interesting and important to understand.

The corpse that had straightforward capital value to a body-snatcher becomes, in the hands of the anatomists, a cadaver which has both 'cultural capital'—the body as a resource for medical knowledge, education and skills—and 'symbolic capital' (Bourdieu 1986)—the body as resource that confers professional legitimization, recognition, honour and prestige. In sum, the transition of corpse to cadaver is a form of translation of capital in the process of commodification.

The procurment of specific kinds of post-mortem human material is driven by the state of medical science and its demands, which changes over time.

The establishment of the NHS in 1948 coincided with altruistic donation with the result that demand and supply levelled out for the first time (Richardson 2006). However, the high commodity value of human material repeats with a difference in the second half of twentieth century as organ donation takes off. By the end of the twentieth century, once organ donation and transplantation surgery become fairly routine, the demand for human organs from recipients needing transplantation surgery outstrips the supply of those willing to donate.

This pattern, to some extent at least, is replicated in pathology where the demand to take organs from the dead for research purposes was voracious, especially at institutions like Alder Hey. During van Velzen's tenure the culture of 'taking every organ in every case' (Redfern 2001a, 2001b) was partly driven by the promise of cultural and symbolic capital gained through research. More surprisingly still, at Alder Hey there were echoes of the earlier culture of body-snatching for straight profit. Alder Hey sold cadavers for five pounds apiece to a pharmaceutical company wanting their pituitary glands in order synthesise human growth hormone (Hurren 2002).

Finally the improper retention of organs was not isolated to the one institution, Alder Hey. Improper retention was a UK wide occurrence (Donaldson 2000). This also resonates with the Georgian period when body-snatching was a widespread phenomenon; and while potentially it threatened all classes, it actually most affected the poor who could ill afford to secure the grave from the body-snatchers.

The Moral Ambivalence of the Collectors of Human Material Over Time
Another historical resonance between body and organ-snatching lies in the morally ambivalent character of medical men such as John Hunter and Dick van Velzen, who were keen to profit from the improper removal of human remains.

John Hunter gained cultural and symbolic capital from the cadaver. Respected in his time, he became a Fellow of the Royal Society in 1767 and today is lauded as being the 'father of modern surgery' and is recognised for his careful observation and scientific method in medicine. As an army surgeon, Hunter contributed to an understanding of gunshot wounds and their early treatment. Post-army, Hunter became an acclaimed pioneer of early transplantation surgery, and he worked on the transplantation of human teeth (Moore 2010).

Hunter is a morally ambivalent character. On the one hand he is lauded for his skills and pioneering work in medicine. On the other hand he can be rightly criticised for his morally questionable methods. Hunter employed agents to obtain stolen human bodies and body parts. He financially induced living donors (often minors) to secure living teeth, which he would implant into wealthy paying adults. Hunter continued his work despite many set-backs where it was clear that he was harming his patients. Perhaps most worryingly, he was blind to such failures and seemingly impervious to ethical criticism (Richardson 2006, p. 159; Moore 2010).

If we look at Dick van Velzen, who became the villain at the heart of retention scandal at Alder Hey, there is scope for comparison.

Both Hunter and van Velzen were prominent social actors in historically conjoined, yet distinctive disciplines. This—as Richardson first pointed out—is relevant by virtue of the fact that pathology evolved from 'morbid anatomy' (Richardson 2006).

Hunter was a practitioner of morbid anatomy and a surgeon, whilst Van Velzen was a professor of paediatric pathology. Like Hunter, van Velzen was actively involved in the improper removal of human material. Hunter paid agents to illicitly procure bodies. Van Velzen directly participated in 'organ theft' by flagrantly abusing consent procedures and lying to parents about the extent of his post-mortems. Again, like Hunter, his motivation seems to have been research led.

Unlike Hunter, van Velzen's research ambitions came to nothing as human materials remained in storage unused. Van Velzen will be remembered as a notorious nonentity. By contrast Hunter is acclaimed as a pioneer because of his contribution to anatomy and 'the birth of modern surgery' (Moore 2010).

Perhaps the most interesting similarity between the two men is their moral blindness. In the transplantation of teeth to live donors, Richardson refers to 'Hunter blindness'; that is, 'the ability to focus so narrowly on recipient benefit as to excise the humanity of the donor from contemplation' (Richardson 2006, p. 159). Van Velzen also seems to have developed a similar form of moral blindness, dogmatically defending his virtue as a researcher and consistently maintaining that he had done 'nothing wrong' in removing and retaining organs from the dead children of distraught parents post-mortem, despite overwhelming evidence to the contrary (Dickson 2001).

Complicated Grief
Richardson states that 'like grave-robbery for dissection, organ procurement necessarily impinges upon the fresh grief of bereaved relatives and friends' (Richardson 2001a, p. 413). However, in her afterword to *Death Dissection and the Destitute* there is no space for further explanation.

Some clues are provided in the parental oral evidence to Redfern (2001b) as to why such posthumous harm was so heartfelt and psychologically damaging. Comparisons may be drawn here to body-snatching in earlier times. In this evidence parents challenge the fact that personal identity ends at death. The mother of Sam, who died at 18 months, puzzles whether Alder Hey understood an 'individual's identity' to end 'at a post-mortem examination, if not death' (Redfern 2001b, p. 425).

For the parents of the child victims of improper retention, the capacity to remember their dead was deeply affected by dismemberment. It is highly likely that body-snatching, some 200 years earlier, would also have complicated grief in a similar way, although there is next to no direct historical evidence surviving of the internal lives of ordinary people to support such a claim.

Public Furore and Parliamentary Intervention
The more evidentially obvious comparison to be made between body-snatching and organ-snatching is in the high-emotion and public furore that both caused.

Bodily theft and respect for the dead generated high emotion and visible public commotion in the Georgian period. For example, catching a body-snatching gang at Lambeth in the district of London in 1795 was reported as:

> "'people of all descriptions, whose relatives had been buried … demanded to dig for them … in great numbers forced their way in, and in spight of every effort the parish Officers could use, began like Mad people to tear up the ground …' 'Great distress and agitation of mind was manifest in every one, and some, in a kind of phrensy, ran away with their coffins of their deceased relations.'" (cited in Richardson 2004, p. 935)

Furthermore, in early January 1832 high emotion led to protest and public violence. A full scale riot erupted at an Anatomy School in Aberdeen, which led to violence, looting, and the school's eventual destruction by being burnt down to the ground (Richardson 2001a, pp. 90–91).

The events at Alder Hey also spawned high emotion and a public furore—though it did not lead to public disorder, violence and the burning down of a medical school. In the case of the organ-snatching, anger was organised through the interest-cum-pressure group PITY2 that spoke truth to power. Parents talked of having their children 'butchered' and 'desecrated' post-mortem without their consent. Many also talked of the hospital having 'betrayed their trust' (Redfern 2001b).

Another interesting comparison between the two cases is how public outrage finally leads to parliamentary intervention. In both cases parliament acted once medical personnel and medical institutions became implicated and incriminated.

An important factor that led from popular protest against the body-snatchers to the select committee that officially looked into the issue was the prosecution and conviction of an anatomist in 1828 for unlawfully conspiring to obtain and receive a body. The proven collusion between an anatomist and resurrectionists 'effectively incriminated anatomical enquiry, and at last caused Parliamentary action to be taken' (Richardson 2001b).

Likewise, once a medical institution was implicated, parliament was forced into a formal investigation of organ retention in the Alder Hey case. Once Ms. Rowlands could no longer reassure parents that the organ retention issue was isolated, insignificant, and media attention forced the issue up the political agenda, both the hospital and van Velzen were incriminated in the ensuing Public Inquiry.

Cultural and Religious Taboo

The final comparison worth drawing is both body-snatching and organ-snatching violated certain cultural and religious taboos.

In the eighteenth century body-snatching violated both cultural and religious taboos by running roughshod over complex funerary customs that existed in caring for the dead. The notion of a 'decent' funeral had strong cultural currency and was deeply entangled in how the dead needed to be treated in order to ensure safe passage in a life hereafter.

With the decline of religion and folkloric belief in the nineteenth and twentieth centuries, comparisons may seem difficult to make. Nevertheless, even in the highly secular times where science supposedly outs belief and superstition about a life hereafter, Alder Hey demonstrates that dismembering the dead was still culturally and religiously taboo.

Many of the Alder Hey parents affected were Catholic and were intent on having a 'decent' funeral, even if it meant multiple funerals in order to bury their children whole. For example, from the parental evidence from the family of Philip (5 years 3 months) to Alder Hey:

> The first funeral was a Catholic burial. The Church was packed with friends, family and work colleagues. Their son should have been buried intact. His body was *desecrated*. The second funeral was very low key. Seven people attended. Their son could not face the second funeral. They feel that the first funeral was *indecent*. They were not sure what they were burying at the second funeral. (Redfern 2001b, pp. 425–426 [author's italics])

The unauthorised removal of Philip's organs was perceived as desecration of his body. This is significant because the word 'desecration' has a double meaning: the harming of 'someone that is dear and loved' and the harming of something that is 'sacred and revered'. It is quite possible therefore, that the desecration of the dead at Alder Hey, in the eyes of Catholic families, deeply contravened their religious belief, as well as complicated grief by affecting their capacity to truly remember them.

Summary

This empirical chapter has historically illustrated posthumous harm and redemption. Posthumous harm and redemption have been understood in three ways.

In contemporary form, posthumous harm has been illustrated through the improper removal, retention and disposal of organs at Alder Hey in the 1990s. This has been interpreted in the following ways:

First, from the formal perspective of the Redfern Inquiry, which provides a broad outline of why and how improper retention can be understood as posthumous harm at the level of procedures and institutions.

Second, from the informal perspective of the parental oral evidence to the Redfern Inquiry, which provides a deep insight into why the improper retention of dead children's organs and tissues mattered so much to parents in the first place.

Third, from the perspective of a historical long view, which compares and contrasts body-snatching with the practise of organ-snatching at Alder Hey.

References

Anonymous. (2001). Return of the body snatchers. *The Economist*. Retrieved May 2, 2016, from http://www.economist.com/node/492949.

Batty, D. (2001, January 30). Alder Hey report on use of children's organs. *The Guardian*. Retrieved February 4, 2016, from http://www.theguardian.com/society/2001/jan/30/health.alderhey1.

Bourdieu, P. (1986). *The forms of capital*. Retrieved May 2, 2016, from https://www.marxists.org/reference/subject/philosophy/works/fr/bourdieu-forms-capital.htm.

Burke, R. (2016). On the Christian burial of the dead. Retrieved May 2, 2016, from https://www.catholicculture.org/culture/library/view.cfm?recnum=3448.

Dickenson, D., & Widdershoven, G. (2001). Ethical Issues in Limb Transplants. *Bioethics, 15*(2), 110–124.

Dickson, N. (2001). Van Velzen interview in full BBC News. Retrieved May 2, 2016, from http://news.bbc.co.uk/2/hi/health/1154181.stm.

Donaldson, L. (2000). *Report of a census of organs and tissues retained by pathology services in England*. London: Stationary Office.

Hall, D. (2001). Reflecting on Redfern: What we can learn from the Alder Hey story? *Archives of Disease in Childhood, 84,* 455–456.

Harrison, J., Innes, R., & Zwanenberg, T. (2003). *Rebuilding trust in healthcare*. Abingdon: Radcliffe Medical Press.

House of Lords Hansard Debates. (2001). *The Alder Hey Inquiry Report 30th Jan 2001*. 621 c. 574. UK Parliament. [Online]. Retrieved February 4, 2016, from http://www.publications.parliament.uk/pa/ld200001/ldhansrd/vo010130/text/10130-06.htm.

Hurren, E. (2002). Patients' rights: From Alder Hey to the Nuremberg Code. *History & Social Policy*. Retrieved May 2, 2016, from http://www.historyandpolicy.org/policy-papers/papers/patients-rights-from-alder-hey-to-the-nuremberg-code.

Klass, D., Silverman, R., & Nickman, S. (1996) *Continuing bonds: New understanding in grief*. New York London: Taylor & Francis.

Moore, W. (2010). *The knife man: Blood, body-snatching and the birth of modern surgery*. London: Bantam Books.

Pilgrim, D., Tomasini, F., & Vassilev, I. (2011). *Examining trust in healthcare—A multi-disciplinary perspective*. Basingstoke: Palgrave.

Redfern, M. (2001a). *The Royal Liverpool children's inquiry report—Summary & recommendations*. London: Stationary Office.

Redfern, M. (2001b). *The Royal Liverpool children's inquiry report*. London: Stationary office.

Richardson, R. (2001a). *Death, dissection and the destitute*. London: Phoenix Press.

Richardson, R. (2001b). Body-snatchers. In C. Blakemore & S. Jennett (Eds.), *The Oxford companion to the body*. Oxford: Oxford University Press. Retrieved May 2, 2006, from http://www.encyclopedia.com/topic/body_snatching.aspx.

Richardson, R. (2004). Bodily theft past and present: A tale of two sermons. *Lancet: Medicine, Crime and Punishment, 364*, 44–45.

Richardson, R. (2006). Human dissection and organ donation: A historical and social background. *Mortality, 11*(2), 151–165.

Sartre, J.-P. (2005). *Being and nothingness: An essay in phenomenological ontology*. (H. Barnes, Trans.). London: Routledge.

Sque, M., Payne, S., & Clark, J. Gift of life or sacrifice? (2006). Key discourses to understanding organ donor families' decision-making. *Mortality, 11*(2), 150–165.

Tomasini, F. (2009). Is post-mortem harm possible? Understanding death, harm and grief. *Bioethics, 23*(8), 441–449.

Open Access This chapter is licensed under the terms of the Creative Commons Attribution 4.0 International License (http://creativecommons.org/licenses/by/4.0/), which permits use, sharing, adaptation, distribution and reproduction in any medium or format, as long as you give appropriate credit to the original author(s) and the source, provide a link to the Creative Commons license and indicate if changes were made.

The images or other third party material in this chapter are included in the chapter's Creative Commons license, unless indicated otherwise in a credit line to the material. If material is not included in the chapter's Creative Commons license and your intended use is not permitted by statutory regulation or exceeds the permitted use, you will need to obtain permission directly from the copyright holder.

Index

A
Accountability, 77
Alder Hey, 73–95
　Alder Hey children's hospital, 3
Anatomy act (1832), 67, 89
Anatomy schools, 90, 93
Anderson (Prof), 74
Annihilation thesis, 25, 27, 29–31
Ante-mortem harm, 21, 24–27, 29, 31, 33
Approach, v
Aristotle, 9
Australian army, 46
Autobiographical death, 2

B
Backwards causation, 26, 27
Bad soldiers, 42
Bare life, 10
Barnett, Corelli, 57
Biological death
　as change, 1, 2; 'Oxford' or real changes, 12, 13, 15; intrinsic changes, 13, 15, 16
　event, 7, 8, 14, 17–19
　process (dying), 7, 8, 11, 17, 19
　state (being dead), 7, 8, 11, 14, 19
Bishop of Liverpool, 88
Bland, Tony, 16
Bloody code, The, 64, 65
Body-snatching, 4, 73, 88–90, 92–95
Bourdieu, Pierre, 34
Brain death
　brain injury, 9, 12, 13
　chronic brain death, 9
　neo-cortical death, 11
　techno-death, 10
　total brain failure, 8, 9, 12
　whole brain death, 8, 11
Brain injury, 2
Bristol Inquiry, 74
British army, 46, 65
British Army Act (1881), 54
British Expeditionary Force (BEF), 47, 53
Browne, Des (Secretary of State for Defence), 48, 57, 60–63

C

Capital punishment, 41, 46, 54, 64, 65, 68, 69
Catholics, 86
Cavan, Lord, 49–51
Christopher (stillborn), 86
Commander-in-Chief (C-in-C), 44
Commodification, 89
Commodity value, 89, 90
Complicated grief, 85, 92, 94
Consent, 73, 74, 76–79, 81, 82, 85, 87, 89, 91, 93
Continuing bond, 85
Corns, Cathryn, 43, 45–49, 51–53, 55, 56, 60–62
Coroner, 75, 77, 80, 82, 83
Corpse, 14
Court martial (CM), 43, 46, 65
Cowardice, 45, 48–50, 52, 58, 63
Cultural capital, 90, 91

D

Dead, 7, 8, 12, 14–16
Death
  as change, 1
  as event
    ; irreversible death events, 8, 12, 16; reversible death events, 12
  normative valence
    ; 'bad' death, 17; 'good' death, 17
  and personal identity
    ; idem identity, 13, 14; ipse identity, 13, 14
Deceit, 81
Deputy Judge Advocate General (DJAG), 43
Desertion, 45, 47, 51–53
Dismembering the dead, 34, 94
Dismemberment, 66, 67
Disrememberment, 66
District Court Martial (DCM), 43
Donaldson, Liam (Prof), 80, 90
Dying, 7, 8, 11, 16, 17

E

Entelelchia, 9
Epicurus, 21–25, 29–31, 33
Existence condition, 15

F

Familial dishonour, 46
Farr, Gertrude, 60
Farr, Harry (Pte), 48–50, 52, 58, 60, 63
Feinberg, Joel, 24–27
Feinberg-Pitcher account, 25–29, 31
Field General Court Martial (FGCM), 43, 44, 48–50
Field punishments, 53
First World War, 3, 42, 46, 53, 55, 62–65, 68, 69
Florida Boy, 8–11
Fogg, Ally, 63
Funerary customs, 73, 79, 94

G

Geatch, Peter, 12
General Court Martial (GCM), 43
Georgian period, 73, 88–90, 93
German army, 46, 65
Gift of life, 87
Good soldiers, 42, 44
Grief, 4

H

Haig, Douglas (C-in-C), 45, 55, 65
Harm, 2, 3

Highgate, James, 26–28
Highgate, Terence, 47
Highgate, Thomas (Pte), 47
Hipkin, John, 48
Historical long-view, 41, 64, 69
Hughes-Wilson, John, 43, 45–49, 51–53, 55, 56, 60–62
Human Tissue Act
 1961, 78
 2004, 88
Hunter, John, 91, 92
Hurren, Elizableth, 67, 69

I
Immortality thesis, 22
Informed consent, 79, 88
Ingham, Albert (Pte), 51–53
Institutional trust, 3, 4
Integrity (of human remains), 86
Intensive Care Unit (ICU), 9, 10, 12
Interdisciplinary, v, vi
Intrinsic harms, 2
 first order intrinsic harms, 32
 second order intrinsic harms, 32

K
Kass, Leon, 10
Kathryn (15yrs), 81
Kenneth (5.5 weeks), 84
King, Peter, 64, 65
Kitchener's volunteers, 47

L
Last judgment, 86
Latham, George (Cpl), 53
Legal death, 68
Life hereafter, 94
Longshore, Alfred (Pte), 51–53

M
Manual of military law, 43, 45, 47, 65
Mark of infamy, The, 35
Meaningful and meaningless lives, 11, 29, 31
Medical death, 8, 9, 12, 14
Medical officer (MO), 42, 44, 49–51
Military justice, 43, 46, 50, 54, 59
Moral blindness, 88, 92
Multidisciplinary, v
Multiple funerals, 94
Murder, 45, 48, 53, 54, 65, 66
Murder act (1751), The, 65, 66

N
Nagel, Thomas, 23
Narrative identity
 configuring personal identity, 14, 15, 17, 19
 refiguring personal identity, 15, 17–19

O
Organ, 73–76, 78, 80, 81, 83, 86, 87, 90, 92, 93
Organ-snatching, 4, 73, 88, 89, 92–94

P
Parental guilt, 85, 87
Parents who inter their young twice (PITY2), 75
Parliamentary intervention, 92, 93
Parris, Matthew, 63, 64
Partridge, Ernest, 24, 25, 31
Passive non-voluntary euthanasia, 16
Paternalism, 78, 79, 85, 88
Peri-mortem existence, 33

Permanent vegetative state, 2, 9, 10, 13, 16
Phantom causation, 27, 28
Philip (5yrs, 3 months), 94
Pitcher, George, 24, 25
Post-mortem procedure, 78, 83
Post sentencing procedure, 43–45, 50–52, 55
Posthumous dishonour, 3
Posthumous harm, 73–75, 79–81, 84, 85, 88, 89, 92, 95
　existence, 22–24, 28, 30–32, 36
　experience, 22–24, 28, 30–34, 36
　harming biography, 28, 33, 34, 36
　harming memory, 28, 31–34, 36
　harming reputation, 29
　impossibility, 21, 29, 31
　symbolic/narrative, 2, 28, 32, 33, 35, 36
Posthumous pardon
　against
　　; exceptionalism, 63; re-writing the past, 56–58, 60; sheer variety of cases, 56, 61; slippery slope argument, 64
　for
　　; orgiveness and reconciliation, 59; historic justice, 57, 59, 61, 62; moral issue, 62; natural justice, 59, 61; revaluating the past, 58–60, 64; symbolic narrative fidelity, 58; symbolic redemption, 59; understanding the past in relation to the present, 60
Posthumous pardoning
　symbolic redemption, 35
Posthumous punishment, 1–3, 41, 47, 64–69
　disrespect to memory, 66
　fear of the afterlife, 66
　as state deterrent, 47, 67
　state power, 34
　as symbolic redemption, 59
　symbolic violence/power, 34, 36
Posthumous redemption, 73, 86, 95
Post-traumatic-stress-disorder (PTSD), 58
Prisoner's friend, 43, 48, 51, 52, 55
Protecting the dead, 85
Punishment, 2, 3
Putowski, Julian, 43–45, 48, 51, 54, 55, 59–61

R
Redfern inquiry (The Royal Liverpool Children's Inquiry Report), 75, 76, 95
Redistributive justice, 47, 65
Regimental court martial (RCM), 43
Regimental discipline, 44, 45, 55, 65
Rehabilitating memory, 3
Religious taboos, 94
Remembering the dead, 2, 34
Resurrection, 86
Resurrection men, 89, 93
Richardson, Ruth, 87–93
Ricoeur, Paul, 13
Rogues, 53
Ross (5 months), 82
Rowlands (Ms), 93
Royal pardon (Royal Prerogative), 63, 64, 68, 69

S
Sachs, Joe, 9
Sam (18 months), 83, 85
Serious incident procedures, 78
Shell shock, 44, 50, 58, 63
Shewmon, 8, 9
Shot at dawn
　shot at dawn policy, 45, 47, 59, 65

Social death
  absolute, 1
  'Cambridge' or relational changes, 12–14
  narrative changes, 13, 15
  processual, 1
Somme campaign, 45, 48, 51, 55
Statutory pardon, 61, 62
Sykes, Julian, 43–45, 51, 54, 55, 59–61
Symbolic capital, 90, 91
Symbolic harms, 84, 85
  second order symbolic harms, 32
Symbolic power/violence, 65, 67

T
Tissue, 74, 78, 82–84, 87
Tony (11 days), 82
Transcendent interests, 26, 28, 29

Trust, 77, 78, 80, 93
Trust in systems, 77
Turing, Alan, 63

U
Uniform determination of Death Act (UDDA), 16

V
Van Velzen, Dick (Prof), 74, 76–78, 81, 87, 88, 90–92, 94

W
War graves commission, 47
Ward, Richard, 64, 65

Open Access This book is licensed under the terms of the Creative Commons Attribution 4.0 International License (http://creativecommons.org/licenses/by/4.0/), which permits use, sharing, adaptation, distribution and reproduction in any medium or format, as long as you give appropriate credit to the original author(s) and the source, provide a link to the Creative Commons license and indicate if changes were made.

The images or other third party material in this book are included in the book's Creative Commons license, unless indicated otherwise in a credit line to the material. If material is not included in the book's Creative Commons license and your intended use is not permitted by statutory regulation or exceeds the permitted use, you will need to obtain permission directly from the copyright holder.

The manufacturer's authorised representative in the EU is Springer Nature Customer Service Centre GmbH, Europaplatz 3, 69115 Heidelberg, Germany. If you have any concerns regarding our products, please contact ProductSafety@springernature.com

Printed and bound by CPI Group (UK) Ltd, Croydon, CR0 4YY
23/03/2026
02076355-0001